So Far From Home

Also by Robert B. Kane

Disobediance and Conspiracy in the German Army, 1918–1945
(with Peter Lowenberg, 2002)

SO FAR FROM HOME

Royal Air Force and Free French Air Force Flight Training at Maxwell and Gunter Fields during World War II

Robert B. Kane

NewSouth Books
Montgomery

NewSouth Books
105 S. Court Street
Montgomery, AL 36104

Published in the United States by NewSouth Books, a division of NewSouth, Inc.,
Montgomery, Alabama.

Publisher's Cataloging-in-Publication data

Kane, Robert B.
So far from home : Royal Air Force and Free French Air Force flight training at Maxwell and Gunter fields during World War II / Robert B. Kane.
p. cm.

ISBN 978-1-60306-369-2 (paperback)
ISBN 978-1-60306-370-8 (ebook)

1. World War II, 1939–1945—Aerial operations, British. 2. Great Britain, Royal Air Force—History—World War II, 1939–1945. 3. World War II, 1939–1945—Aerial operations, French. 4. World War II, 1939–1945—Aerial operations, American. 5. Aeronautics, Military—Study and teaching—Great Britain. 6. Aeronautics, Military—Study and teaching—Alabama. 7. History, Military—20th century.
I. Title.

2016956404

Design by Randall Williams and Erin Mills

Printed in the United States of America
by McNaughton & Gunn

In Memoriam, Oakwood Cemetery, Montgomery, Alabama
To the memory of the 78 Royal Air Force (above) and 20 Free French (below) flight cadets who died during their respective flight training programs at the flight schools of the Southeast Air Corps Training Center/Eastern Flying Training Command, June 1941 to November 1945. They died in a foreign land far from home, learning to fly so they could return home and fight for the freedom of their homelands during World War II.

Contents

Acknowledgments

As the current director of history for Air University at Maxwell AFB, Alabama, it was a privilege for me to research and write this book about the flight training of Royal Air Force (RAF) and the Free French Air Force (FFAF) in central Alabama during World War II and dedicate it the 78 RAF cadets and the 20 FFAF cadets who died in training accidents in the southeast United States so far from home. I have visited the graves of these young men and am always moved by their sacrifice and the solemnity of their eternal resting place. I would like to acknowledge the following individuals and organizations that made this achievement possible. Foremost is my wife Anita who endured late night and weekend studies as I worked on the book. I also thank now retired Air Force Colonel David M. Cohen, former Director of Staff for Air University and my immediate supervisor, who allowed me some time to conduct some of the needed research in the Air Force Historical Research Agency (AFHRA) at Maxwell Air Force Base, Alabama, which holds the official histories and original documents that provided most of the material for this paper. Finally, I must provide a special thanks to the staff of the AFHRA in general and Mr. Sylvester Jackson, archivist at the AFHRA who found the histories and documents, used in this study, among the vast holdings of the Agency and made them available for my use, and Ms. Tammy Horton and Mr. Carl Bailey, who provided the pictures from AFHRA official histories, used in this book. Finally, I thank the staff, especially Ms. Josephine Turner, Interlibrary Loan, and Mr. Michael Rojas, Circulation Staff, of the Muir S. Fairchild Research and Information Center (MSFRIC) (Air University Library) at Maxwell AFB.

— R. K.

Abbreviations Used

AAC	Army Air Corps
AAF	Army Air Forces/Army Airfield
ACTS	Air Corps Tactical School
AFB	Air Force Base
AFHRA	Air Force Historical Research Agency
AETC	Air Education and Training Command
AVM	Air Vice Marshal
BCATP	British Commonwealth Air Training Plan
BFTS	British Flying Training School
CIS	Central Instructors School
DoD	Department of Defense
ECATP	Empire Commonwealth Air Training Plan
EFTC	Eastern Flying Training Command
ENJJPT	European-NATO Joint Jet Pilot Training
FFAF	Free French Air Force
FTC	Flying Training Command
GAF	German Air Force
GC	Group Captain
HQ	Headquarters
IOS	International Officer School
LAC	Leading Aircraftsman
NATO	North Atlantic Treaty Organization
n. d.	no date
RAF	Royal Air Force
RTC	Replacement Training Center
RTU	Replacement Training Unit
SEACTC	Southeast Air Corps Training Center
TOEFL	Test of English as a Foreign Language
US	United States
US AAF	United States Army Air Forces

Foreign Ranks and AAF Equivalents

Royal Air Force (RAF) Ranks and Army Air Force Equivalents

Raf Ranks	*Rank*	*Army Air Force Equivalents*
Air Marshal	O-8	Major General
Air Vice Marshal	O-7	Brigadier General
Air Commodore	O-6	Colonel
Group Captain	O-5	Lieutenant Colonel
Wing Commander	O-4	Major
Leading Aircraftsman	E-2	Private

Free French Air Force Ranks and Army Air Force Equivalents

In general, the FFAF ranks mentioned in this narrative were equivalent to those of the AAF; for example, a FFAF *lieutenant* was equivalent to an AAF first lieutenant (O-2), and a FFAF *captaine* was equivalent to an AAF captain (O-3). Among the French flight cadets who came to the United States during World War II for AAF flight training were *aspirants*. The *aspirant* was not a commissioned, warrant, or noncommissioned officer. In the 1940s, the closest American equivalent was an officer cadet, attending the US Military Academy (West Point) or the US Naval Academy (Annapolis). The French military services used the rank to designate officers in training at French military academies or voluntaries, serving as temporary officers. As the narrative will demonstrate, the French officers and enlisted men treated the *aspirants* as if they were officers, and the *aspirants* saw themselves as "junior officers" who wanted the privileges and courtesies of officers. After going through his training and a probationary period, the *aspirant* would be commissioned as a *sous-lieutenant* (Army or Air Force second lieutenant) or *enseigne de vaisseau de deuxième classe* (Navy ensign) (O-1). It was widely used during both world wars to provide young educated men with the authority of an officer.

List of Tables

So Far From Home

Introduction

The history of the world has many examples in which the military forces of one country provided formal training to the military forces of its friends and allies to enhance relations with their friends and allies and to improve their military capabilities as one means to augment their own military power in case of conflict with an enemy country. The history of the United States (US) provides several examples of mutual military training to improve the quality of American military forces and those of its friends and allies. For example, members of the British Army trained the organized militias of the British Atlantic colonies in North America before 1770. A Prussian officer, Baron Friedrich Wilhelm von Steuben, trained soldiers of the Continental Army during the American Revolution. After the US government declared war on Germany in April 1917, the US Air Service pilots who went to France learned to fly French and British aircraft in operational flying schools in France. After war clouds again appeared over Europe (and also the Pacific), Germany had occupied most of Europe, including France, and Britain stood alone against the Nazi war machine twenty years later, the United States returned the favor by training British and French flight crew in the United States.

During World War II, the United States Army Air Forces (USAAF), starting in the summer of 1941 before the formal entry of the United States into the war, provided flight training to over 21,000 members of 29 Allied countries (Appendix 1) by late 1945, funded by the Lend-Lease Act of March 11, 1941 (Appendix 2). The two largest programs provided flight training to over 12,000 members of the Royal Air Force (RAF) and over 4,000 members of the Free French Air Force (FFAF). Of these, about 6,000 RAF and FFAF aircrew members went through various phases of flight training at Maxwell and Gunter fields near Montgomery, Alabama, under the Southeast Air Corps Training Center (SEACTC, later Eastern Flying Training Command, or EFTC) by late 1945.[1] Because of ongoing German air attacks on Britain, overcast weather, the need to use all available airfields for the defense of Britain, and the occupation of France by Nazi Germany in June 1941, Britain had limited facilities, airfields, instructors, and aircraft to conduct flight training for new flight crews for the RAF. Later, after the Allies cleared North Africa of German and Italian forces, the Free French government-in-exile, headed by General Charles de Gaulle, asked the United States to train pilots for its nascent air force. Although the pilots, trained in central Alabama during the war provided a relatively small percentage of the aircrew who flew with the RAF and FFAF in Europe and North Africa by the war's end, this training positively impacted inter-Allied relations and foreshadowed the far more extensive flight training that the US Air Force would conduct for allied air forces from 1947 to the present.[2]

In the huge library about World War II and the US Army Air Forces, no one (as far as I can tell) has yet written the complete story about the flight training for the AAF or allied air forces during World War II. The vast majority of the secondary literature about American and Allied air forces during that war focuses on aerial operations, combat operations, combat aircraft, operational units at various levels, and various individuals, mostly senior officers, of the AAF and the RAF. The only secondary work on AAF training of Allied air crews that I found was a monograph, *Training of Foreign Nationals by the AAF 1939–1945*, by Gerald T. White, published in 1947 by the AAF Historical Office, and the most "extensive" work on AAF training during World War II are several chapters in *The Army Air Forces in World War II, Vol. VI, Men and Planes* by Wesley F. Craven and James Lea Cate, which has a section on Allied flight training during the war.[3] More specifically, *The Arnold Scheme British Pilots The American South and Allies' Daring Plan*, published in 2008 by Gilbert S. Guinn, is the only extensive study of the AAF training of RAF flight cadets in the Southeast immediately before and to mid-1943. In addition, to my knowledge, there are no secondary works on Allied flight training in central Alabama during the war or on the flight training of FFAF aircrew during the war, at least in English.[4]

As a result, this study on the AAF training of RAF and FFAF flight cadets, the two largest allied flight training programs of World War II, in central Alabama will fill a current gap in the historical literature of World War II in general and the AAF training of allied air forces during the war in the United States in particular. In addition to providing some idea of the impact of flight training of RAF and FFAF flight crews at Maxwell and Gunter Fields on the Allied conduct of World War II, this study will also provide the impact on American-Allied relations during the war as well as its influence on the flight training provided by the US Air Force to pilots of allied air forces, especially members of the North Atlantic Treaty Organization (NATO) after 1947.

1

Lend-Lease Funding of Allied Flight Training

On September 1, 1939, the German Army invaded Poland, precipitating the start of World War II in Europe. Two days after the German invasion of Poland, Britain and France declared war on Germany. By May 1940, France had deployed a significant portion of its army of five million men into the Maginot Line, and Britain had dispatched almost 395,000 soldiers and airmen to northern France.[5]

After six months of "phony war" or "*Sitzkreig*" ("sit-down war"), German troops occupied Norway, Denmark, The Netherlands, Luxembourg, and Belgium; defeated France; and drove the British from the continent between April 9 and June 22, 1940. Between July 10 and September 7, 1940, the *Luftwaffe* (German Air Force) came dangerously close to defeating RAF Fighter Command as the condition for a seaborne invasion of Britain, but a series of mistakes and wrong decisions by the *Luftwaffe* High Command and Adolf Hitler, especially the switch to bombing British cities, led Hitler to first postpone and then eventually cancel the projected invasion and gave Fighter Command a needed respite to reconstitute itself. Now standing alone against Nazi-occupied Europe, Britain looked to the Dominions and the United States for help, especially the latter for war materiel and, eventually, flight training.

By late 1939, British government and the RAF leaders realized that the RAF could not train large numbers of flight crew in Britain under current wartime conditions and looked to the Dominions and the United States to provide flight training for the needed aircrew. The German air attacks on Britain and the start of RAF strategic bombing of Germany required the RAF to use its available airfields, aircraft, and experienced pilots for ongoing combat operations. Furthermore, British weather conditions with overcast skies and fog, sometimes for days, further limited the available good flying weather and daylight hours for training and operational flying.[6]

Years before, in 1936, the British government and the RAF leadership had opened discussions with the governments of the Dominions to establish a massive flight training program for the air forces of the British Commonwealth. However, because of the Depression and reluctance by these governments, little was done until October 1939 after the war had started. The discussions about a Commonwealth-wide flight training program proceeded rapidly, and, on December 17, 1939, representatives of Canada, Australia, New Zealand, Southern Rhodesia, South Africa, and India signed the final plan. Called the Empire Commonwealth Air Training Plan (ECATP; after June 1942, renamed as British Commonwealth Air Training Plan, BCATP), the plan envisioned training tens of thousands of flight crew over time in flight schools located in a number of the Dominions. Work began on the training schools in early 1940, and the first students began arriving at the first ECATP school in April 1940.[7]

Meanwhile, U.S. President Franklin D. Roosevelt, believing that a Nazi-dominated Europe threatened the national security of the United States, realized that, at some point in the future, the United States would have to help Britain and eventually become involved in the war in Europe. However, with 8.1 million Americans still unemployed (14.6 percent unemployment rate) and significant isolationist and non-interventionist sentiment in the country, including the US Congress, Roosevelt had to go slowly and incrementally to convince the American people and Congressmen of the danger of a Nazi-dominated Europe and the need to help an isolated Britain.[8]

Although German and Italian aggression in Europe after September 1939 slowly reduced the level of isolationist and non-interventionist sentiment in the United States, their adherents were still present, influential and vocal, and the Neutrality Laws of 1935, 1936, 1937, and 1939 further limited Roosevelt's ability to legally help Britain in its time of need.[9] As a result, Roosevelt preceded slowly in the name of US national security to help Britain and other victims of Nazi aggression short of war.

Following the November 1940 elections in which Roosevelt won an

1.1. Harry Hopkins, special assistant to President Franklin Roosevelt, leaves the US for Britain on January 6, 1941, to discuss with British government officials specific requirements for US military aid to Britain.

unprecedented third term, he continued to receive news of the deteriorating condition of Britain, including that Britain was running out of cash to purchase American war materiel, required by the "cash and carry" provisions of the November 4, 1939, Neutrality Act.[10] This dire financial situation came to a head when Philip Kerr, Lord Lothian, British ambassador to the United States, told US newsmen at LaGuardia Field on November 23, "Well boys, Britain's broke; it's your money we want." The comment angered Roosevelt who believed Lothian was trying to force him to ask Congress to repeal the "cash and carry" provision, but Roosevelt was not ready to confront the isolationists and non-interventionists in Congress. On December 9, Roosevelt received a letter from Winston Churchill, Chamberlain's successor as British prime minister, in which Churchill asked Roosevelt for warships, especially destroyers; an extension eastward of the US self-declared defense line; and financial relief from the "cash and carry" provision of the 1939 Neutrality Law. From this letter, Roosevelt and Harry Hopkins, a close advisor to Roosevelt and one of the architects of the New Deal, generated the basic concept of lend-lease: lending Britain war materiel and allowing Britain to pay for it after the war.[11]

At a press conference on December 17, Roosevelt explained the concept of lending or leasing war materiel to Britain, his well-known "garden hose" analogy. On the evening of December 29, Roosevelt gave his "arsenal of democracy" fireside chat, which so moved public opinion that a Gallup Poll in early January 1941 showed that 68 percent of Americans favored lend-lease.[12]

Roosevelt now wanted a first-hand assessment of Britain's current defense and financial situation and of its new prime minister, Winston Churchill, but, at this particular moment, neither country had an ambassador from the other country as Lord Lothian had died suddenly on December 12, 1940, and Joseph P. Kennedy Sr., the US ambassador to Britain, had resigned in late November. After several delays during which Hopkins helped Roosevelt develop his Lend-Lease concept and draft his upcoming State of the Union address and inaugural address, Roosevelt announced on January 3, 1941, that Hopkins would be going to Britain to assess the ability of Britain to eventually prevail over the Germans with American aid and convey to Churchill that the United States would help the British in their fight against the Nazis.[13]

Hopkins left for Britain on January 6, 1941, and arrived on January 9. He spent six weeks, traveling around Britain, observing bomb damage from German air raids, viewing coastal defense fortifications, and meeting with Churchill. Churchill in these meetings let Hopkins know what Britain needed to defeat Hitler, and Hopkins reaffirmed America's resolve to help Britain defeat Hitler. He also met with British experts on finance, industrial production, and shipping and sent Roosevelt numerous cables that detailed the war materiel that Britain needed. Before leaving Britain on February 9, Hopkins helped Churchill prepare a radio speech on the need for American aid, particularly aimed at American as well as British audiences.[14] In his radio broadcast later that evening, Churchill assured his American listeners, "We shall not falter; we shall not weaken or tire. Neither the sudden shock of battle, nor the long-drawn trials of vigilance and exertion will wear us down. Give us the tools and we will finish the job."[15]

On January 6, 1941, Roosevelt gave his State of the Union address in which he asked Congress for the authority and funds to manufacture additional weapons and war materiel to give to those countries fighting the Axis Powers.[16] Between January 6 and January 9, Secretary of the Treasury Henry Morgenthau, General Counsel Edward H. Foley, and others crafted the bill that Senator Alben W. Barkley (D-KY) and Representative John McCormack (D-MA) introduced in their respective Houses on January 10. Debate on the Lend-Lease bill now began in the US Congress while Hopkins was still in Britain with opposition to the bill from the America First Committee, founded at Yale University by R. Douglas Stuart Jr., a Yale law student, in September 1940. This committee included a number of prominent Americans as members or supporters, such as Alice Roosevelt Longworth, the daughter of Theodore Roosevelt; George N. Peek, the former head of the Agricultural Adjustment Administration; Eddie Rickenbacker, the World War I ace; Lillian Gish, the silent film actress; Henry Ford for several months; and Charles A. Lindbergh, the celebrated aviator.[17]

Despite the anti-bill testimonies, the House passed the bill with a vote of 260–165 on Hopkins' last day in Britain. Right after Hopkins returned

1.2. President Franklin D. Roosevelt. On March 11, 1941, he signed the Lend-Lease Act.

to Washington, the Senate began its debates on the Lend-Lease bill. Again, despite intense anti-Lend-Lease rhetoric, the bill passed the Senate on March 11 by a vote of 60–31. The House passed the Senate-amended bill by a wide bipartisan margin of 317–71 on March 11, and Roosevelt signed the bill into law the same day.[18]

The Lend-Lease Act (Appendix 2) was not the first attempt of the US government to use US education and training facilities to influence foreign governments, contribute to better relations with them, and contribute to US national security. In many ways it was a global extension of the Goodwill Act, signed by Roosevelt on June 28, 1938. Related to his Good Neighbor Policy to improve the image of the United States in Latin America, the act allowed a limited number of Latin Americans to attend US government professional educational institutions and schools. Executive Order 7964, issued by Roosevelt on August 29, 1938, directed the executive departments that maintained educational institutions to develop regulations to govern foreign training.[19]

The President also directed each department to report the number of students that each school could accept without reducing the enrollment of American students, provide the entrance requirements for foreign students, and the provisions to protect nation defense information. Those Latin American countries which desired to send students to US government schools had to submit applications to the Secretary of State who, after consulting with the respective department, would grant the request. In reality, most countries dealt directly with the War Department. While the act had real training goals, it also had political goals: future collaboration of US military services with those of Latin American countries, eventual elimination of the influence of fascist ideology in Latin America, and influence on future Latin American leaders to favor the United States. Although very few Latin Americans received training in US military schools before the war began, the Goodwill Act served as the legal basis for much of the training provided to Latin Americans before, during and after World War II.[20]

The passage of the Lend-Lease Act solved the problem of providing war material and defense services, including aircrew training, to Britain and other countries and gave the President the discretion he needed to allocate war material to countries which needed it. The new law allowed the President to "sell, transfer title to, exchange, lease, lend, or otherwise dispose of" any munitions or defense-related items to "any country whose defense the President deems vital to the defense of the United States."[21]

However, critics believed that the President had received a "blank check" with virtually no limits. Hopkins notified Churchill about the passage of the bill, and Churchill replied, "The strain had been serious. I thank God for your news," and later touted it as "a new Magna Carta." Roosevelt quickly obtained an initial appropriation of $7 billion to fund the initial war material purchases, appointed Hopkins to administer the aid program, and established the organization and staffing for the lend-lease program.[22]

2

RAF Flight Training in the US

While the Congress debated the pros and cons of the Lend-Lease bill, Major General Henry "Hap" Arnold, the chief of the Army Air Corps (AAC),[23] and his staff examined the British recommendations (requests) for aircraft to bolster Britain's defense against the *Luftwaffe* that Hopkins had communicated to the President. These requests included 50 PBY Catalina amphibious aircraft with equipment, munitions, and spare parts; engines for 20 engineless Lockheed aircraft already in Britain; 100 P-40 Tomahawk aircraft with machine guns; the maximum number of B-17 bombers; 200 AT-6 Texans (called "Harvards" by the RAF) or BT-13 Valiant trainers in addition to ones already delivered; five additional fully equipped civilian flying schools; and a plan to ferry bombers to Britain.[24] After General George C. Marshall, Army Chief of Staff, reviewed the requests, he commented to Arnold on February 27, "We are going to be forced to some very radical thing or departure." One Army officer commented, "We can't take seriously requests that come late in the evening over a bottle of port." On March 6, Arnold informed Marshall that filling the British requests would adversely his efforts to create a viable combat force for the AAC. In concluding his comments, Arnold recommended the creation and dispatch of a joint military mission to Britain to determine how best to meet Britain's requirements and continue building the AAC.[25] The question now became how could the US government meet both objectives of building the AAC into a modern and effective air force while also helping Britain in building the RAF into an effective air force in the face of regular *Luftwaffe* air attacks against Britain.

As Congress debated the Lend-Lease bill and Marshall and Arnold wondered how the AAC could accomplish these objectives, the British government established four schools for "refresher" training for American

2.1. General Henry H. "Hap" Arnold, Chief of the Air Corps, Sep 29, 1938–Jun 19, 1941, and Chief of the Army Air Forces, Jun 20, 1941–Feb 28, 1946. He is the only American to have been a five-star general in two military services (US Army, Chief of the AAF, Dec 21, 1944; and the US Air Force, May 7, 1949), and the only five-star USAF general.

volunteers, recruited by the Clayton Knight Committee for the RAF and the Royal Canadian Air Force (RCAF), with British funding, the result of discussions with the State Department in June and August 1940.[26] Between fall 1940 and March 1941, the British Air Commission had contracted with four American flying schools, the Spartan School of Aeronautics at Tulsa, Oklahoma; the Dallas School of Aviation at Dallas, Texas; the Polaris Flight Academy at Glendale, California; and the Joe Ploesser's Aeronautical Training Center, Inc., Bakersfield, California, paid with British funds. The contracts also included the purchase of aircraft to conduct the training. The original purpose was to "grade" current American volunteers and provide refresher training to those whose licenses had expired or those who were close to attaining the required 300 flying hours. These pilots would then make their way to Canada for operational posting in Britain. Later, these schools trained pilots who had little or no flying experience.[27] However, by March 1941, the British had run out of money for more aircraft and to operate the fourth school. At this juncture, Arnold at a March 7 meeting with representatives of six civilian flying schools and the British Air Attaché in Washington, Air Commodore Sir George C. Pirie, offered to release up to 260 primary and 285 advanced trainers in the next three months for these flying schools through Lend-Lease. The British Air Commission subsequently closed these initial contractor schools in 1942, apparently as the contracted BFTS and the Arnold Plan (see below) got underway.[28]

Meanwhile, in light of Arnold's views and the President's concerns, Hopkins on March 10 recommended to Marshall that he send a contingent of American military leaders, specifically including Arnold, to Britain to validate the British requests with their British counterparts. Later in the month, Secretary of War Henry L. Stimson discussed the mission to Britain with Assistant Secretary of War John J. McCloy and Special Assistant for Air Affairs Robert A. Lovett. By March 25, all had agreed that Arnold and others would go to Britain to discuss Britain's "shopping list" for American war materiel. Over the next two weeks, cables crossed each other across the Atlantic with suggestions by Hopkins and British officials on what Arnold should see and do, including an audience with King George VI and the British Foreign Secretary Anthony Eden. By the time Arnold left on April

2.2. Before the US entry into World War II, about 9,000 Americans had joined the Royal Canadian Air Force. After the US declaration of war on Dec 7, 1941, many returned to the United States to complete their flight training and wartime service with the American military. These four, standing in front of a North American AT-6 Texan, were photographed at Maxwell Field's advanced flight school before August 1942 when the SEATC closed the school.

9, 1941, the main objectives of his visit were to determine the numbers and types of US aircraft to be produced, their allocation among the various claimants for the aircraft, and the potential for the United States to train British pilots and American crews to ferry aircraft to Britain.[29]

Arnold and his aide, then Major Elwood "Pete" Quesada, left New York City on April 10, 1941, and arrived in Britain via Bermuda, the Azores, and Lisbon, Portugal. During his three-week visit, Arnold met his RAF counterpart, Air Marshal Charles Portal; the American special envoy to Britain, Averell Harriman; the Minster of Aircraft Production Lord Beaverbrook (William "Max" Aitken); Chief of the Imperial General Staff, General

Sir John Dill; King George VI; the British Foreign Secretary; and many other British and American officials. He even met Lord Hugh Trenchard, the British airpower theorist and first chief of the RAF, the world's first independent air force. Arnold visited bombed cities, coastal fortifications, antiaircraft defenses, defense industries, and RAF installations across Britain. He received briefings on Britain's current defense status and estimates of German military capabilities and met a number of RAF combat pilots. During the course of the discussions Arnold offered to provide a third of the AAC's pilot training capacity to meet the RAF's need for trained pilots. After a busy and full three weeks, Arnold and his entourage left Britain for the United States on April 27 and arrived in New York City through Lisbon, West Africa, Brazil, and the British West Indies on May 1.[30]

Arnold's conversations with Air Vice Marshal (AVM) John C. Slessor, the RAF Director of Plans; AVM Sir Charles E. Medhurst, Assistant Chief of the Air Staff (Intelligence); and AVM Sir Alfred Guy R. Garrod, Air Member for Training, on April 13 concerned the specific requests, related to British airpower needs, that Hopkins had brought back to Washington, earlier in the year. At this meeting, Arnold offered to provide one-third of AAC training facilities and establishments to train RAF pilots, British usage of the US navigation school at Miami, Florida, provision of American pilots to ferry aircraft to Britain, and ferry routes from the United States to Britain. Later conversations with AVM Sir Willifrid Freeman, assistant to Portal, covered problems with the existing ferry service, prioritization of American aircraft types for production, and additional heavy bombers.[31] According to Arnold, he "had every reason to believe we would be able to get some sort of an agreement on these subjects before my departure."[32] Arnold's offer of training facilities in the United States apparently had a great impact on the British air force leaders. According to Colonel Raymond E. Lee, the US Military Attaché in Britain, they were "completely staggered by this offer which so far exceeded all their expectations." Churchill in a letter to President Roosevelt called the proposal "splendid, unexpected and a very welcome addition."[33] This discussion contained the origin of the Arnold Plan to provide US Army Air Corps (AAC) facilities, including those in Central Alabama, to train large numbers of RAF pilots.[34]

With the complete acceptance of Arnold's proposal to open AAC (Army Air Forces after June 20, 1941) facilities to the RAF, Marshall and Arnold now needed to turn the proposal into reality. The British had first brought up the idea of training RAF aircrew in the United States in September 1940 during the Battle of Britain (see earlier narrative on the British refresher schools), and Hopkins had also brought back the proposal for training facilities from his trip in Britain in early 1941. In March, Roosevelt directed Arnold to examine the proposal, and Arnold, in turn, discussed the proposal with the operators of six civilian flying schools in February before Arnold had left on his trip to Britain. After Arnold's return, General Marshall sought authorization to fund the RAF training program with Lend-Lease funds. With this funding authorization, Arnold recalled the directors of the civilian flying schools to discuss the use of their schools to train RAF pilots, assisted by RAF observers. At the same time, Arnold arranged for the Navigation School at Coral Gables, Florida, operated by Pan American Airways, to train RAF navigators.[35] On May 20, 1941, President Roosevelt approved these programs to train RAF flying cadets in the United States.[36]

Meanwhile, in late April, a question came up about the legality of using Lend-Lease funds to provide replacement aircraft for Army aircraft transferred from current Army flying schools to the British contractor schools and to provide flight training for the RAF that brought planning for RAF flight training to a standstill. The former issue came about because in a December 23, 1940, letter to the Secretary of War Stimson Roosevelt prohibited the use of government funds to replace aircraft, spare aircraft engines, spare parts, and so on, ostensibly needed for US defense needs, which the Army would have to turn over to the British so they could begin flight training at the six US contractor schools. Without the removal of this paragraph from the letter, the Army could not legally obtain reimbursement for these aircraft. Oscar S. Cox, who helped draft the original Lend-Lease Act as Assistant to the General Counsel, US Treasury Department (1938–April 1941), and now General Counsel of both the Lend-Lease Administration and the Office of Emergency Management (April 1941–1943), worked with Robert H. Jackson, US Attorney General, to develop a legal solution to these problems. In early May, Hopkins notified Lovett that the President's letter

2.3–4. Left: Air Chief Marshal Sir George Pirie, British Air Attaché to the United States, Oct 1, 1937 to Aug 13, 1941. Right: Air Marshal Alfred Guy R. Garrod, Member of Air Council for Training, Jul 8, 1940 to Apr 27, 1943.

would be modified so that the Army could obtain replacement aircraft for the aircraft made available to the contractor schools.[37]

The second issue, funding the 4,000-pilot program (see below), developed apparently on another technicality—"training" was not a weapon, a piece of equipment, a munition, or defense information as specified in Section 2 of the Lend-Lease Act. In a general sense, the Attorney General answered these and other similar questions that came up from Lend-Lease detractors in a March 27, 1941, address prepared by Jackson and read to international lawyers and jurists at the first meeting of the Inter-American Bar Association in Havana, Cuba. Generally, Jackson took a "liberal" reading of the growing and evolving body of international law and called the aggression of the Nazis, Fascists, and militarists

> . . . civil wars against the international community. Accordingly, as responsible members of that community, we can treat victims of aggression in the same way we treat legitimate governments when there is civil

> strife and a state of insurgency—that is to say, we are permitted to give to defending governments all the aid we choose.

As a result, given that these aggressors were menacing "the political, territorial, economic and cultural integrity of the Western Hemisphere," the United States was justified in giving aid to the Allies short of war.[38]

On this matter, Attorney General Jackson stated that the President had the legal authority under Lend-Lease and the Constitutional authority as Commander-in-Chief to make flight training available to foreign nations.[39] On May 17, in response to a question from the President about legally funding the AAF training of RAF cadets the day before, Robert P. Patterson, Under Secretary of War, informed FDR that the Justice Department assured him that "there [were] no legal obstacles [to fund] the program."[40] Probably earlier, but certainly by early 1943, "defense articles" and "defense information," two types of items, authorized by the Lend-Lease Act, became defined as "all types of goods and services necessary for the waging of total war." Lend-Lease aid by late 1942 consisted of "goods transferred," meaning physical items such as weapons, and "services rendered," services needed to provide the physical items to Allies or to train Allied military members in their use. As a result, the various programs to train pilots of Allied countries in the US fell under the category of "services rendered," such as shipping and supply services, the development and operation of ferry routes and supply bases overseas, and production facilities in the United States to manufacture war goods and, thus, legally eligible for funding under the Lend-Lease Act.[41]

By summer 1941, as a result of various discussions, the AAF and the RAF had developed three programs to train RAF aircrew in the United States.[42] The smallest program was the training of RAF navigators at the Pan American Navigation School at Coral Gables, Florida, established on August 10, 1940, to train 100 AAC navigator cadets in long-range aerial navigation techniques, many of which Pan American had developed, every six weeks. Early in 1941, Arnold had offered the RAF ten spaces in each class, and the first cadets from Canada entered training on March 24, 1941. By early summer, Pan American increased the number of spaces per class to 200 with RAF cadets filling 150 seats, beginning on July 7, and Americans filling the remaining 50 seats. As result, this program could provide up to a thousand RAF navigators per year. In July 1942, Headquarters AAF withdrew the 150 RAF quotas from the Navigation School, and, beginning on August 28, 1942, Americans filled these former RAF quotas. The last class of RAF navigators graduated from the school on October 17, 1942, bringing the total of RAF navigator graduates of this program to 1,177 over the past 19 months.[43]

The second program was known as the British Flying Training School (BFTS) program and sometimes as the "six-week school plan" and the "3,000-pilot program." It involved the training of 50 RAF cadets per class at five-week intervals for a 20-week course. The classes would be held at six new civilian flying schools which would be established by the six civilian contractors currently training AAC flying cadets. In addition, this program would provide a refresher course for at least 18 students for an eleven-week course.[44] For this program, the AAC only helped the RAF and the contractors select sites for the school and supervised the construction of the schools and, then, when the schools were in operation, assisted the RAF in logistical issues, including supply and maintenance, and ancillary issues, such as medical care. The British government furnished $3 million for the construction of the schools, and the US government provided the land through Lend-Lease. The RAF had complete control over the actual training program at these schools. Because the contractors had not completed the schools by June 7, 1941, the scheduled date for the first 200 RAF cadets to begin training, they entered four of the existing schools in groups of 50 cadets each as overloads. When the cadets for the third class arrived in the United States on August 23, five of the schools were open: at Mesa, Arizona; Lancaster, California; Clewiston, Florida; Miami, Oklahoma; and Terrell, Texas. The sixth school at Ponca City, Oklahoma, opened soon afterwards. A seventh school operated at Avenger Field, Sweetwater, Texas, for a short time in the summer of 1942. (This airfield later became the training site for the Women Airforce Service Pilots [WASP] from February 1943 to December 1944.)[45] This program had graduated 6,921 pilots for the RAF when the last of the schools closed on September 11, 1945.[46]

The third program was the training of RAF flight cadets at various AAF installations alongside American flight training cadets. While the RAF, AAF, and the civilian flying school contractors worked out the details of the BFTS program, Arnold, as stated earlier, offered one-third of his pilot training capacity to the RAF. He believed that the AAF schools could train 4,000 pilots annually at an estimated cost of $33,706,330 of Lend-Lease funds.[47] Arnold selected the Southeast Air Corps Training Center, headquartered at Maxwell Field near Montgomery, Alabama, to oversee that program, known as the "4,000-pilot plan" or "Arnold Plan." The RAF cadets would receive the same training as the American cadets at 12 schools, four of which were operated by contractors and the remaining eight directly by the AAF, as noted in Table 2.1.

Table 2.1.

AAF Schools for the Arnold Plan

Primary Schools

Alabama Institute of Aeronautics, Inc., Tuscaloosa, Alabama*
Carlstrom Field, Arcadia, Florida
Darr Aeronautics Tech, Inc, Albany, Georgia*
Graham Aviation, Co. Souther Field, Americus, Georgia*
Lakeland School of Aeronautics, Lakeland, Florida*
*AAF Contractor Schools

Basic Schools

Cochran Field, Macon, Georgia
Gunter Field, Montgomery, Alabama

Advanced Schools

Craig Field, Selma, Alabama
Maxwell Field, Montgomery, Alabama
Napier Field, Dothan, Alabama
Moody Field, Valdosta, Georgia
Turner Field, Albany, Georgia

The AAF flight schools provided ten weeks of instruction, each, in three phases of flight training, primary, basic, and advanced. In the fall of 1941, the AAF added five weeks of preflight instruction. (See Appendix 4, Phases of AAF Flight Training, c. 1943–44, for specifics about each phase.) During the duration of the Arnold program, June 7, 1941 to March 17, 1943, 4,370 RAF cadets graduated out of the 7,860 who entered the program, a 55.6 per cent graduation rate.[48]

By the summer of 1941, the AAF and the Navy had established four programs to train RAF pilots (Appendix 2). Although the numbers of flight crew who graduated from the American training programs were small in comparison to the numbers of flight crew that the BCATP would eventually produce, they represented the willingness of the US government to initially assist Britain in holding back the Germans and taking the war to the German people at the time when Britain stood alone in Europe against Nazi Germany. Additionally, the US government made this initial significant commitment at a time when it was still legally neutral. The speed at which the US government and the AAF established these flight training programs for the RAF demonstrated the commitment of the US government and the AAF leadership to help Britain short of war and also hinted at the tremendous potential of the US government, its military forces, and its renewed industrial production to aid countries around the world in their fight against the aggressors of 1941. By fall 1941, it was apparent that a growing percentage of the American population and Congressmen were behind the President to help Britain and other countries against Germany, Italy, and Japan. However, some detractors later argued that the President's actions of 1941 with the passage of the Lend-Lease Act and subsequent aid in various forms to the beleaguered countries of the world caused the aggressors to initiate more aggressive and provocative actions against the United States that eventually led to the attack on Pearl Harbor on December 7, 1941, the declaration of war of the United States on Japan on December 8, and the subsequent declaration of war by Hitler against the United States on December 11.

Above: 3.1. Aerial photograph of Maxwell Field, circa 1938, before the War Department constructed numerous barracks for the students of the AAF flight schools and permanent party at Maxwell. Below: 3.2. Aerial view of Gunter Field, 1942. Inset: 3.3. Emblem of the SEACTC.

3

Royal Air Force Flight Training in Central Alabama

On June 25, 1941, RAF Aircraftsman E. J. Briggs of Worcester, England, reported to the AAF advanced school at Maxwell Field from the primary school at Tuscaloosa, Alabama. Unfortunately, he was rushed to the post hospital for an emergency appendectomy and spent his first few days at Maxwell in the hospital. After coming out of anesthesia, he asked, "How long before I can get back to flying?"[49] So began the flight training of more than 7,800 RAF flight cadets who passed through AAF flight schools in 1941–43. The first 550 cadets were scheduled to arrive on June 7, but many incurred immigration delays as they entered the United States from Toronto, Canada.[50] On October 19, 1941, Wing Commander Henry A. V. Hogan presented a Union Jack to Captain J. R. Luper, commandant of cadets, Maxwell Field Replacement Center, as a symbol of Anglo-American solidarity.[51] By February 1943, when the RAF and the AAF terminated the AAF training of RAF cadets, over 7,000 had arrived in the United States for flight training, and over 4,000 had received their wings.

Maxwell Field near Montgomery, the state capital of Alabama, had a long association with flight. In early March 1910, the Wright Brothers established the first civilian flight school in the United States on a nearby cotton plantation. Because of high winds and the time it took to receive parts from the Wrights' workshop in Dayton, Ohio, the school closed after two months of operation. In July 1918, the War Department opened Aircraft and Engine Repair Depot #3 at the same location as the earlier Wrights' flight school. In 1922, the field was named after Second Lieutenant William C. Maxwell, an Alabama native who had died in a crash in the Philippines in August 1920. The airfield not only remained active during the budget-conscious 1920s, but it continued to grow through the lobbying efforts

3.4. Staff of the Southeast Air Corps Training Center (SEATC), Maxwell Field, in front of what is now Building 800, Austin Hall, the Air University headquarters at Maxwell AFB. Between September 1931 and summer 1940, the building housed the Air Corps Tactical School. Brig. Gen. Walter R. Weaver is in the front row, fifth from the left.

3.5. Maj. Gen. Walter R. Weaver commanded SEATC from Aug. 20, 1940, to December 1941. Weaver had served as Maxwell Field post commander 1927–31, and commandant of the Air Corps Tactical School 1939–40. He subsequently served as acting chief of the Army Air Corps and as commander of Army Air Forces Technical Training Command.

of Alabama Congressman J. Lister Hill. In 1928, the War Department's decided to relocate the Air Corps Tactical School (ACTS) from Langley Field, Virginia, to Maxwell Field. The school opened in September 1931 and taught airpower tactics and developed airpower doctrine—especially high altitude, daylight precision bombing doctrine—until spring 1940.[52]

With war clouds gathering in Europe and Asia, the War Department began expanding US military forces, which included the training of 7,000 pilots and 3,600 bombardiers and navigators annually. The Air Corps' plans meant terminating the education mission of the ACTS at Maxwell and turning Maxwell into a training installation. During a May 1940 visit, General Arnold addressed Maxwell Field personnel and informed them and Colonel Walter R. Weaver, the post commander and the commandant of the ACTS, about the Air Corps' flight training plan and the conversion of Maxwell into one of four flying training centers. On July 8, 1940, the AAC established the Southeast Air Corps Training Center (SEACTC), headquartered at Maxwell Field, to oversee flight and technical training in the Southeast United States, with now Brigadier General Weaver as its first commander. The AAC also established a basic flying school at Maxwell Field that became operational on September 14, 1940, and an advanced single-engine school that began on November 20, 1940. Over the next several years the AAF would establish schools for preflight, basic, primary and advanced (twin- and four-engine) flying training, and navigator and flexible gunnery training throughout the Southeast and Midwest United States under the SEACTC (Eastern Flying Training Command after July 31, 1943). By September 1945, SEACTC had conducted flight training for one-third of the AAF pilots, bombardiers, and navigators during the war.[53]

In late May 1940, along with the notification to establish SEACTC, General Arnold also directed the establishment of a basic flying training school (phase 2 of the AAC's flight training program), originally slated to open with 75 students in September 1940 at Maxwell Field. A week later, he notified Colonel Weaver that he had decided that the basic school would be located "in the vicinity of Montgomery," rather than at Maxwell, and directed Weaver to locate a suitable site with existing suitable facilities. Weaver recommended that the War Department acquire the Montgomery

3.6. Aerial view of Maxwell Field, mid-1942, after construction of barracks for the thousands of Americans and internationals who trained there during World War II. Compare to photograph 3.1.

municipal airport, four miles northeast of the city, and began negotiating with the city, county, and state to acquire the property. Alabama Governor Frank M. Dixon agreed to lease the airport for $1 per year for 35 years. The War Department drew up the required contract with an effective transfer date of July 1, 1940. On August 27, 1940, Weaver officially activated Gunter Field, named after Montgomery's recently deceased Mayor William A. Gunter, and approved the start of needed renovations to current structures, the construction of other facilities, and the search for nearby auxiliary fields for cadet flight training. By December 1941, Gunter Field had 108 flight instructors, 108 Vultee BT-13 trainers, and most of its needed facilities.[54]

SEACTC received official notification that it would have a major role in the RAF pilot training program in April 1941, initially for the training of RAF cadets at its civilian contractor schools. In early May, Major William W. Welsh, Weaver's executive officer, returned from Washington, DC, with news that the first 550 students would enter six SEACTC primary schools on June 7. Although General Weaver received little additional information about the "Arnold program" in the next several weeks, he did receive guidance for the RAF cadets who would be trained in the contractor schools. The RAF cadets would hold the rank of aircraftsmen (enlisted pilots) and would not be subject to US Army regulations or Articles of War (the 1940s version of today's Uniform Code of Military Justice). However, AAF officers could issue orders to the RAF cadets, and they would observe American military courtesy and customs. Also, the upperclassmen would oversee the underclassmen. The RAF cadets would receive training, using the same curricula as the American cadets, meet a flying board when they were having problems in training, take medical exams, and participate in physical fitness training, like the Americans. Instructors would provide reports on each trainee, just as with the Americans.[55]

To administer the RAF flight training programs (the contract schools, the Arnold Program, and the Pan American navigation school) in the United States, the RAF in July 1941 established the RAF Delegation in Washington. The delegation consisted of the chief and several additional RAF officers attached to Arnold's staff at HQ AAF to deal with the assignment and movement of RAF cadets from Canada to flying schools in the United States and back to Canada after completion of their training programs and acquisition of aircraft and equipment for the RAF. The chief of the delegation would coordinate with the RAF liaison office in Ottawa, Canada, where RAF students arrived and departed through a reception center. Additionally, the RAF would assign an officer to the staff of the headquarters of each of the AAF training centers and also to each of the AAF schools which had RAF students to assist with local training issues, involving RAF cadets, to include disciplinary actions of the RAF cadets in consultation with local AAF authorities.[56] AVM Arthur T. Harris was the first chief of the RAF Delegation serving from May 27, 1941, to February 21, 1942, when he returned to Britain as chief of RAF Bomber Command.[57]

RAF cadets for the Arnold training program arrived at Halifax, Nova Scotia, from Britain, usually aboard former ocean-going passenger liners, which, since they were faster than German U-boats, sailed without escort vessels and as single ships. From Halifax, the cadets of the first five classes went by train to No. 1 Personnel Depot (PD) 1, Toronto, Ontario, for immigration checks and medical inspections before entering the United States. Students for the sixth through 11th (and last) classes entered from No. 31 Personnel Depot (No. 31 PD), Moncton, New Brunswick, completed in October 1941.[58] Cadets for Classes 42-A, B, and C were issued a light blue and dark blue cloth badge with the letters "U.K." to distinguish them from the American cadets and a duffel bag with a complete winter flying outfit, goggles, and flying boots, then proceeded in civilian clothes by train to their primary schools. After "in-processing," the cadets took the train to Maxwell Field. To better prepare incoming RAF cadets, Generals Arnold, Weaver, and Brett in June 1941 approved sending 1,000 trainees to the Replacement Training Center (RTC), which opened at Maxwell Field August 9, 1941. The RTC consisted of two schools, the Air Corps Replacement Center (Pilot) for flying cadets and the Air Corps Replacement Center (Air Crew) for flying cadets who failed pilot training or were selected as bombardiers and navigators.[59] Following graduation from the RTC, the cadets went to their various primary training schools within SEACTC.[60]

Group Captain (GC) David V. Carnegie, the British Air Attaché and Director of British Flying Training in Washington, DC, 1941–September 11, 1942,[61] approved the RTC curriculum "enthusiastically" and thought "the program was excellent in every respect." However, he recommended reducing the number of hours on military law from 11 to 6 and administration time from 10 to 5, and eliminating the 16 hours set aside for vaccinations, which RAF cadets had already received in Canada. He also recommended adding American geography, aircraft, and terminology for these hours, resulting in the following program:

Table 3.1

Program of Instruction for RAF Cadets in the Maxwell Replacement Training Center/Pre-Flight School, September 1941– January 1942

Subjects	Hours
Reception and Processing	17
Academic Preparation	
American Geography and History	4
Citizenship	2
Mathematics	20
Terminology and Outline of Elementary Flying Training	4
Recognition of American Airplane Types	4
Military Hygiene and First Aid	5
Chemical Warfare Defense	2
Current Events	2
Administrative Indoctrination	
Customs and Courtesies of Service	5
Organization Lectures	5
Basic Military Indoctrination	6
Physical Training	48
Miscellaneous (Pay Periods, etc.)	2
Total	**159**

Source: Section III, Chapter Exhibit 1, AAF EFTC History 7 Jan 1939–7 Dec 1941, AFHRA Call Number 222.01 7 Jan 1939–7 Dec 1941 v. 2, Document # 001416134. Also see Maxwell Field History 7 Dec 1941 – 31 Dec 1942, Vol. 1, pp. 210–37, AFHRA Call Number 286.26–2 v. 1, Doc # 00175352, for a detailed narrative of the RTU/Pre-Flight School curriculum late 1941/early 1942.

RAF cadets in Classes 42-E (750), 42-F (750 of which 5 were eliminated), and 42-G (750) attended the RTC course at Maxwell Field, but, because of congestion, subsequent classes, starting with Class 42-G, attended the RTC course at the exclusively RAF school at Turner Field, Albany, Georgia. Graduates of the Maxwell Field RTC and subsequent classes of the Turner Field RTC then entered one of the primary training schools. Graduates of the primary schools then went on to one of the basic schools, including the one at Gunter Field.[62] See Appendix 4 for details about the phases of the AAF flight training program and Appendix 5 for the

3.7. British flight cadets in dark shorts and American flight cadets in khaki shorts conduct joint physical fitness training on the Maxwell Field parade ground just to the west of the Headquarters SEATC building. Today, Lister Hall, Building 804, which houses the offices of the 42nd Air Base Wing, sits on the former parade ground.

trainers used in each phase of flight training.

The basic school, operating at Gunter Field since September 1940, received its first RAF cadets in September 1941, Class 42-A. The 10-week program included 70 hours of flying instruction in taxiing, take-offs, landings, stall, turns, spiral, and acrobatics, and other air maneuvers, utilizing the Vultee BT-13. In the first few months of the school's operations, half the class flew in the mornings and the other half in the afternoons. By the summer of 1941, Gunter had increased the number of instructors and training aircraft and acquired several auxiliary fields in the Montgomery area that allowed the school to train more cadets safely each day. By then, the ground school portion of the school consisted of 98 hours: radio code (30 hours), air navigation (15 hours), meteorology (30 hours), radio communications

Table 3.2.

RAF Statistics for the Gunter Basic School

Class	Start Date *	RAF Entered	Amer Entered	RAF Held Over	Amer Held Over	RAF Eliminated	Amer Eliminated	RAF Killed	Amer Killed	RAF Graduated	% RAF Grad.	Amer Graduated	% Amer Grad.
42-A	8/18/41	158		2				0		149	95.5		
42-B	10/4/41	155		1		6		1		149	95.5		
42-C	11/4/41	204		0		19		0		186	90.7		
42-D	12/16/41	197		5		36		0		156	81.3		
42-E	1/10/42	240		4		41		2		198	82.2		
42-F	2/24/42	175	72	10	0	21	8	3	0	145	85.8	64	88.9
42-G	3/29/42	205	77	7	0	31	9	8	0	169	82.4	65	84.4
42-H	4/30/42	209	86	3	2	25	6	3	0	185	86.9	78	92.9
42-I	6/2/42	148	176	0	3	24	27	0	1	127	84.1	147	84.0
42-J	8/31/42	143	142	0	2	10	11	2	0	131	91.6	165	92.3
42-K	8/6/42	189	121	8	4	15	11	3	0	163	90.1	108	90.8
43-A	9/8/42	184	172	1	1	14	3	0	0	177	92.7	172	98.3
43-B	10/10/42	184	158	3	2	13	2	0	1	169	92.9	154	98.1
43-C	11/15/42	3	364	0	11	0	20	0	4	6	100	331	93.2
Totals		2,387	1,368	44	25	249	97	22	6	2,114	88.6	1,284	93.9

*Graduation dates for the respective classes were: 42-A, 11/11/41; 42-B, 12/3/41; 42-C, 1/9/42; 42-D, 2/20/42; 42-E, 3/17/42; 42-F, 4/29/42; 42-G, n/a; 42-H, Jul 5/42; 42-I, 8/6/42; 42-J, n/a; 42-K, 10/10/42; 43-A, 11/11/42; 43-B, n/a; 43-C, 1/26/43.

Note: data not found for Americans in classes 42-A through 42-E.

Source: For RAF trainees: AAFSEATC Pilot Class History, Classes 42-A – Class 43-C, United Kingdom Trainees, AAF EFTC History, 7 Dec 1941-1 Jan 1943, Appendix, Supplemental Exhibits, Chapters 9, 27, 39, AFHRA Folder Call # AAF EFTC 7 Dec 1942-1 Jan 1943, Doc # 00146140; For American trainees: AAFSEATC Pilot Class History, Classes 42-F – Class 43-C, Total Trainees, AAF EFTC History, 7 Dec 1941-1 Jan 1943, Vol. 4, AFHRA Call # AAF EFTC 7 Dec 1942-1 Jan 1943 v. 4, Doc # 00146138.

3.8. RAF cadets return to their barracks after graduation exercises from the Maxwell Field advanced flight school. Photograph taken before August 1942 when SEATC closed the school.

(3 hours) and airplane and engine operation (20 hours). Given the differences between the Americans and the RAF cadets, the school revised the curriculum by Class 42-C, increasing navigation from 15 to 20 hours and reducing radio code from 30 to 25 hours. The first Gunter basic school class with RAF cadets was Class 42-A, starting in August 1941 and consisting of 168 RAF cadets and 30 Americans. See Table 3.2 for statistics for the Gunter basic schools.[63]

Class 42-A through 43-C of the Gunter basic school graduated 2,181 Americans (49.3 percent of graduates) and 2,144 RAF cadets (50.7 percent), saw the elimination of 156 Americans (39.4 percent of those eliminated) and 249 RAF cadets (60.6 percent of those eliminated), and the death of five Americans (20 percent of deaths) and 22 RAF cadets (80 percent).

In August 1941, SEACTC officially designated the advanced school at Maxwell as a twin-engine school but, because of a shortage of twin-engine trainers and equipment, it operated as a single-engine school throughout its existence. As a result, the students in Classes 42-A, B, C, D, and F, almost entirely of RAF cadets, entered the Maxwell Advanced School in 1942 and received training in single-engine aircraft, despite constant reminders about revisions in the course of instruction in the twin-engine school program. The actual and intended course of instruction introduced ambiguity into the conduct of the school. The school also lacked current textbooks and adequate training films on various topics, such as gunnery, formation flying and combat flying. Normally, cadets in the single-engine advanced training went to Eglin Field for fixed gunnery training, but SEACTC and RAF officials canceled this training for the RAF cadets in Classes 42-C, D, E, and F as they were destined for duty as bomber pilots. Additionally, the school suffered from insufficient instructors to maintain the stated five students to one instructor ratio; some instructors had six or seven students. The advanced school at Maxwell closed with the graduation of Class 42-F on July 3, 1942 as HQ SEACTC in July 1941 had decided to move the school to Turner Field, Albany, Georgia, possibly to make way for the Central Instructors School (CIS).[64]

Table 3.3

RAF Statistics for the Maxwell Advanced School

Class	Began	Size	Held Over	Killed	Elim-nated	No./% Graduated	Grad-uated
42-A	11/11/41	68		4	0	64/94.1	1/3/42
42-B	12/15/41	70	1	0	1	72/98.6	2/7/42
42-C	1/11/42	73	0	0	0	73/100	3/6/42
42-D	2/20/42	78	1	0	0	77/100	4/24/42
42-E	3/17/42	80	6	1	3	71/94.7	5/17/42
42-F	5/1/42	77	2	1	4	76/93.8	7/3/42
Totals		446	10	6	8	433/97.1	

Source: AAFSEATC Pilot Class History, Classes 42-A – Class 42-F, United Kingdom Trainees, *AAF EFTC History, 7 Dec 1941–1 Jan 1943*, Appendix, Supplemental Exhibits, Chapters 9, 27, 39, AFHRA Folder Call # AAF EFTC 7 Dec 1942–1 Jan 1943, Doc # 00146140.

As the RAF flight training program got underway, the AAF began selecting graduates of the advanced schools to serve as instructors in the primary and basic schools. Since mid-1940 the AAF had begun ramping up its own flight training program, and it faced a shortage of flying instructors at all schools. It had determined that the preferred instructor-student ratio in the basic and advanced single-engine schools was one instructor to five students and in the advanced twin-engine schools, one instructor to 6.7 students. To approach this ideal ratio, the AAF began recruiting civilian instructors from early 1942 through mid-1943 and offered commissions after a very brief training program at Officer Candidate School to obtain volunteers.[65]

Given the shortage of instructors for the American flight training schools, the AAF and the RAF began selecting RAF graduates of the advanced schools for instructor duty. Those selected to receive commissions went to No. 1 Manning Depot, Toronto, Canada, obtained RAF uniforms, and RAF pilot officer wings, and then went to Maxwell Field for the instructor's course. For example, 42 graduates of Class 42-F, 60 from Class 42-G, 88 from Class 42-H, 60 from Class 42-I, 37 from Class 42-J, 24 from Class 42-K, 46 from Class 43-A, and a number of graduates from Class 43-B were retained and sent to the five-week program at the Central Instructors School at Maxwell Field, officially activated on May 15, 1942.[66] Most of these RAF instructors went to the basic and advanced schools which trained RAF cadets.[67] For example, by the end of 1942, the Gunter basic school had 55 RAF instructors, 46 of whom had graduated from SEACTC schools, nine from the CIS, and 14 from both.[68] Victor Hewes, a RAF cadet who completed his flight training through the AAF schools and returned to Maxwell in September 1941 for the instructor's course, stated that the

> US instructor's school was nothing compared to the RAF FIS (Flying Instructor School) course. All we did was fly around in the back seat of the twins. At no time were we ever told how to instruct, or even how to explain the maneuvers (sic) we were doing. They had nothing similar to the RAF patter system. We just had to demonstrate our ability to fly the various maneuvers (sic) from the instructor's seat.[69]

Although a tour as an instructor pilot delayed their return to home and family in Britain and entry into combat, it also provided the airmen with more flying hours, additional pay, and freedom from the blackouts and rationing in Britain, and it might have saved their lives, as casualties were high among multi-engine RAF pilots in the European theater.[70] By November

1942, when SEACTC terminated this program—deemed "wholly satisfactorily" and "a great success," according to qualified SEACTC officers—it had graduated 457 RAF flying instructors for various SEACTC schools. With the termination of the RAF flight training with the graduation of Class 43-C in late February 1943, SEACTC replaced the RAF instructor pilots with Americans by July 1943.[71]

To administer the British flying cadets at the various SEACTC schools, GC Carnegie, the British Air Attaché and director of British Flying Training, assigned a senior RAF officer to Weaver's staff at Headquarters SEACTC; the first was Wing Commander Henry A. V. Hogan, June 1941–August 1942, who had responsibility for all matters related to the administration, discipline, welfare, and training of the RAF cadets. He sat on any board concerning an RAF cadet, provided his "expert" opinion on any issue related to RAF cadet training, and sent frequent reports to the British Air Attaché. Hogan also commanded young or young-middle-aged RAF flying officers assigned to the different SEACTC schools to monitor the progress of RAF cadets. These RAF officers, including those at Maxwell and Gunter Fields, rated each RAF cadet according to his qualifications for a commission and his aptitude for flying fighters or bombers. The forms used for these ratings became an important part of the cadets' records which RAF officials in Britain used to determine commissions (only one-third were commissioned) and what type of aircraft a cadet would fly after returning to Britain, especially since most American instructors tended to rate most RAF cadets as "average" with few details.[72] The RAF officer at a SEACTC school also supervised any RAF flight instructor at the school after SEACTC began using RAF advanced school graduates as instructors.[73]

By January 1942, the AAF training of RAF flight cadets was more or less operating smoothly. The RAF cadets entered the United States from Canada and proceeded by train to the RTC at Turner Field, Georgia, for in-processing and basic military education and training. After completing the RTC course, they went to one of several primary schools for nine weeks of flight training. Graduates then went to one of several basic schools, such as the one at Gunter Field, for the second nine-week flight training phase. By the time of graduation from the basic school, the RAF Delegation had decided which students would attend an advanced single-engine school, such as the one at Maxwell Field, or an advanced twin-engine school. Initially, the AAF advanced schools were short on twin-engine trainers, and, as a result, RAF graduates of the basic school received training in single-engine trainers, contrary to the RAF Delegation's wishes, as the RAF wanted the graduates to fly bombers. Carnegie wrote Weaver about the RAF's desire to provide RAF cadets with all possible twin-engine training. Despite Carnegie's request, RAF students did not receive twin-engine aircraft training until spring 1942, and SEACTC sent the older RAF instructors of the basic schools to the twin-engine schools training RAF cadets.[74]

About one month before the end of the advanced training school, the school would provide the No. 31 PD with the number of RAF students graduating from that course. At the same time, the school would forward the names of RAF cadets—no more than a third of graduates—that the school commander recommended for a commission, based primarily on the cadet's potential as an officer and, secondarily, his proficiency as a pilot. SEACTC then radioed the names (confirmed later by letter) to the RAF Delegation in Washington for review and confirmation by an RAF Commissioning Board. The RAF Delegation would then authorize the commissions, subject to final approval by the Air Ministry. The advanced schools would prepare the diplomas for graduation and forward them to Headquarters SEACTC for signature by the SEACTC commander. Those approved for commissions would be promoted to Pilot Officer and the remaining graduates would be promoted to the rank of Sergeant Pilot. Graduates would then proceed to No. 31 PD to receive appropriate uniform insignia before returning to Britain.[75] (See Appendix 6 for detailed information about the RAF training program from Class 42-A through Class 43-B.)

During training the RAF cadets had few serious problems. The RAF cadets had initial difficulty in understanding the American language and terminology, and many worried about their families in Britain.[76] Also, they disliked the AAF discipline, derived from the Aviation Cadet Handbook, which was "considerably more strict" than in Canada or Britain. The presence of an American upper class which lorded over the lower class of RAF cadets in Class 42-A led to hazing and British objections. Many found many practices of the AAF training system—marching at the double, calling senior cadets "Sir," demerits for petty offenses, no talking at meals, eating "square meals"—"schoolboyish and unnecessary," according to John Evans, another "Arnold" cadet. These practices received so much criticism from American cadets as well as RAF cadets that the AAF Training Command abolished them in 1943.[77] American officers found the behavior of some RAF cadets "rebellious." Subsequent classes had a number of combat veterans who had seen action in France and Belgium in 1940 and were not pleased with American disciplinary ideas. For example, a commander at Gunter reported that RAF cadets refused to attend lectures. In another situation, some cadets apparently had cheated on examinations, and the school authorities decided to use different tests for each subsequent class to prevent cheating.[78] An interesting situation occurred with the first class of RAF cadets at the Gunter school. Nine had managed to retain their moustaches, "their pride and joy," according to one cadet, throughout the primary school at Craig Field, but the commander of the Gunter School ordered the cadets to comply with AAF requirements to be "clean shaven." In October, after hearing complaints from RAF cadets, General Weaver allowed the RAF to retain their moustaches as long as they did not interfere (interestingly enough) with eating.[79] Overall, although the RAF Director of Personnel received an occasional letter from Gunter RAF cadets, complaining of training center conditions, these complaints "did not reflect the general attitude of the Royal Air Force trainees or indicate an undesirable state of morale."[80]

At the start of the RAF training program, RAF officials had estimated a 20 percent elimination rate from the primary schools, which proved to be unrealistic.[81] However, by the end of the last class, 43-B, which graduated on February 26, 1943, 4,360 of the 7,862 cadets who had entered the 13 classes between June 1941 and March 1943 made it through the complete flying program, a 55.5 percent completion rate (the remaining 44.5 percent included cadets who "washed out" for flying deficiencies, the largest percentage; were terminated for other reasons, such as illness and disciplinary actions; and died from flying and nonflying accidents).[82] Despite the official wishes of the RAF and the AAC leaders to maintain high standards during the flight training programs, word-of-mouth indicated that instructors were more interested in training as many RAF pilots as possible and thus retained cadets who should have been eliminated.[83] Actual student achievement, however, seemed to contradict this belief. Many RAF cadets, such as Victor Hewes, believed that they were "washed out for no good reason except a personality conflict with the instructor or some minor infraction of the regulations." Other RAF cadets believed that their reactions to the American style of discipline contributed to a good number of eliminations.[84]

The average RAF student soloed after 10 to 12 hours, with most eliminations occurring before that time. Overall, the number of RAF cadets completing the full training program increased through Class 42-C, decreased from Class 42-D through Class 42-F, and rose again with Class 42-G, ending with over two-thirds of the last class, Class 43-B, completing the program. Generally, the percentage of cadets completing the first phase steadily increased from 55 percent to 82 percent. Those completing the primary school fared much better in completing the basic and advanced schools.

Table 3.4

SEACTC Schools Elimination Rate, June 1941–March 1943

Phase of Training	American (%)	British (%)
Primary	30.1	35.6
Basic	7.7	6.3
Advanced	1.7	0.8
Overall	**39.5**	**42.7**

Source: "Comparative Analysis of Elimination Rates among Aviation Cadets (Americans), U.K. Students and Student Officers Since Class 42-A," Section III General Tab, *AAF EFTC History, 7 Dec 1941 – 1 Jan 1943*, Vol. 4, AFHRA Call # 221.01 7 Dec 1941 – 1 Jan 1943 v. 4, Doc #00146138.

Overall, the elimination rates of the American and the RAF students for flying deficiencies between June 1941 and March 1942 (Classes 42-A to 42-F) were about the same. The RAF cadets did worse than the Americans in the primary schools, possibly because they did not initially attend a pre-flight school, but performed slightly better in the basic and advanced schools.

Given the larger than expected number of eliminees, SEACTC and the

RAF Delegation had to develop procedures to accommodate them while they waited for transportation to Canada.[85] In August 1941, General Arnold approved General Weaver's recommendation to allow 75 of the 242 RAF cadets eliminated from Class 42-A to reenter the primary school because of the haste in assembling the cadets for training and possible problems with adjustment to the training regimen. GC Carnegie objected to this decision. Eventually, only 23 of these students entered Class 42-C of whom 22 completed the training.[86]

To occupy the eliminees so they would not desert, SEACTC gave them academic work in the bombardier-navigator pool at Maxwell. However, there were actually 180 eliminees, instead of 50 to 75, and there was a problem with funds to feed them. In September 1941, SEACTC directed the RAF action officers at the schools to send the eliminees to Canada for ground service or training in non-flying specialties.[87]

Of greater concern was the high accident rate of the RAF trainees, one fatality out of a hundred—a total of 78 by February 1943 (9 in the primary schools, 33 in the basic schools, and 36 in the advanced school). Overall, in terms of flying hours, the RAF cadets, despite beliefs by some Americans, fared about the same as the American cadets. In April 1942, the only month for which complete comparative statistics were available, there were 538 training accidents, 183 British and 355 American. The basic schools had 37 British accidents and 33 American; and the advanced schools had 53 British and 66 American. For the entire RAF training program, the RAF cadets in the primary schools had 6.5 accidents per thousand flying hours while the Americans had 5.1. In the basic schools, RAF cadets had 1.34 accidents per thousand flying hours, and the Americans had 0.6. In the advanced schools, the RAF cadets had 1.96 accidents per thousand flying hours while the American cadets had 2.0.[88] In general, the number of accidents per thousand flying hours was not significantly different between the Americans and the RAF cadets in the first three phases of the AAF training program.

SEACTC attributed the "high" accident rate of the RAF cadets to several causes. Some believed that the RAF cadets were dangerous flyers: buzzed highways at low altitudes, causing car accidents; flew under bridges;

conducted mock dogfights; and bounced their aircraft in front of vehicles on the highway. Some believed that the civilian and Army instructors were lowering their standards to graduate as many RAF cadets as possible. Still others believed that the RAF cadets lacked mechanical skills supposedly common to American boys and thus had greater difficulty in flying an airplane, resulting in more accidents. Other noted causes included poorer eyesight and the mental attitudes of the students. Finally, observers believed that the general dislike of the foreign students for American flying regulations led them "to dispense as much as possible with discipline in the air."[89]

During the entire RAF training program the worst series of accidents occurred on May 20, 1942, when seven RAF cadets died when their aircraft crashed in bad weather. A flight of aircraft with 34 students, nine instructors, and a training squadron commander left Gunter Field on a training flight to Crestview, Florida, with a planned return to Gunter via Mobile, Alabama, that night. After some discussion about the weather, the flight took off and reached Crestview safely. By then, the weather had deteriorated. The students took off from Crestview, but the flight leader could not contact the others as the weather worsened. Some students tried to return to Mobile, others tried to land, and the remainder tried to fly around the bad weather. In the end, seven students were killed when their aircraft crashed, others sustained minor injuries, and others landed safely all over southern Alabama. While no one was severely punished, the weatherman at Gunter Field was relieved from forecasting duty, the acting director of flying training was transferred to another base, and HQ SEACTC prohibited all night training.[90] By the end of 1942, 20 RAF cadets had died in training accidents, compared to five American fatalities for the entire year.[91]

A 1944 official AAF study cited these as the problems for "high level" of RAF training accidents. However, the report did not mention that the high level of RAF accidents occurred at the highest level of RAF participation in the training program. Nor did the report mention that training accidents involving American cadets showed some of the same problems as the RAF cadets after the AAF dropped the two-year college requirement, lowered physical standards, and abolished the upper class–lower class system and the hazing of underclassmen, the latter over General Weaver's protest. Finally,

3.9. Lord Halifax, British ambassador to the United States, visits the RAF graves in Montgomery's Oakwood Cemetery in 1943.

according to reports from Britain about the performance of RAF pilots from the AAF training schools in combat, resistance to American discipline during their training had not produced special problems from the graduates who flew RAF night bombing raids over Germany, "which required firm discipline as well as stout courage."[92]

On November 15, 1941, RAF cadets Frank V. Marhoff and Richard N. Moss, the first Maxwell cadets killed when their aircraft collided in midair two days before, joined several RAF cadets killed in training accidents from other schools since June in Oakwood Cemetery in north Montgomery. Earlier, WC Hogan had purchased a section of the cemetery for the purpose of interring RAF cadets killed in accidents. Each cadet was buried with full military honors as fellow cadets bore the Union Jack-draped coffin to the burial site, with a contingent of American cadets following. An AAF chaplain conducted the service, and a bugler played "Last Post," the British equivalent of "Taps."[93] The SEACTC commander ensured that photographs of the funeral procession, the services, and the gravesite covered with flowers were sent to the bereaved families in Britain. The school commander apparently received numerous letters from family members thanking him, the school, and the cadets for the sensitivity and the honors of the funeral services.[94] Olive Shotbolt, the mother of RAF cadet Robert A. Shotbolt of Class 42-F, wrote,

> . . . I am sure [the funeral] could not have been carried out any better, and he was buried just as my husband and I and members of my family wished. I hope that you will thank all Officers and Cadets that followed him and also those sending wreaths. I am very grateful to everyone who saw him to his last resting place, also the U.S. Army Air Corps who provided the firing squad. . . . [H]e died for his Country which is a big honour and was laid to rest with full honours, for which once again I am very grateful to you all.[95]

A letter from Mrs. Mary E. Clare expressed similar feelings about the funeral of her son, T. W. Atkins, also from Class 42-F.[96]

The RAF cadets at Maxwell and Gunter Fields drew considerable interest from the local Montgomery population. The SEACTC intelligence officer had directed the public affairs office not to provide information about the RAF cadets to the local media, probably because the United States was still neutral. Yet, the local newspapers managed to pick up sufficient information about the RAF cadets to publish stories about their training program.[97] The RAF cadets could not have automobiles and could not accept social invitations that interfered with their training. Still, they met locals at the USO clubs, churches, and other local venues and received invitations to dances, parties, and private homes; some even managed to travel around the eastern United States, time permitting.[98] For example, the student council and International Relations Club of Huntingdon College hosted a tea on October 19, 1941, for a group of RAF cadets, and Mrs. Henry C. DeBardeleben of Lowndesboro, about 25 miles southwest of Montgomery, invited 50 RAF cadets from Gunter Field and American cadets from Maxwell Field to her plantation in nearby White Hall.[99]

3.10. Member of the RAF Delegation saluting the RAF cadets buried at Oakwood Cemetery, Montgomery, Alabama.

That same month, ladies of Montgomery's St. John's Episcopal Church, along with Mrs. Venetia Hogan, the wife of WC Hogan, formed the British War Relief Society which raised funds from local businessmen and civic organizations to help provide for the needs of the British and American flight cadets at the nearby airfields. As more RAF cadets arrived for training, the Society decided to sponsor "teas and biscuits" every Saturday at St. John's.[100] Starting Sunday, October 19, 1941, Gunter and Maxwell Fields had open

3.11. RAF flight cadets from Maxwell and Gunter Fields march in the Nov 1941 Armistice Day parade in downtown Montgomery to the cheers of the Montgomery area citizens watching the parade.

houses or parades in which Montgomerians viewed the American and RAF cadets marching with the American and British flags in the color guard. They compared the Americans' "relatively motionless and easy stride" with the Britishers' "sharp resounding high steps and stiff, high swinging arms." The RAF contingent at the November 11, 1941, Armistice Day parade in Montgomery received the largest volume of applause.[101] The RAF cadets also held a British-style dance, featuring British novelty dances, such as the "Palais Glide" and the "Lambeth Step," with SEACTC leaders as guests and local girls as dance partners.[102] In December 1941, 1,200 American and British cadets ate Christmas dinner in Montgomery homes.[103] Overall, the local population held the RAF cadets in high esteem.

Setting the tone of the RAF flight training program in central Alabama, General Weaver commented, at a lunch hosted by the Montgomery Chamber of Commerce,

> It's a happy sight to see marching side-by-side down Montgomery's streets today a British soldier and an American soldier, while flying over them both the Red, White and Blue as well as the Union Jack. That speaks to me of a solidarity of the English-speaking peoples, which should be consoling in these times.[104]

RAF cadets did have some interesting reactions to Southern culture, Montgomery, and Montgomerians. They were not used to the sultry Southern climate and the greasy Southern cooking.[105] In downtown Montgomery, they were more impressed with coin-operated soda machines than the historic buildings, given the cadets' familiarity with much older historic buildings in Britain. There were concerns about understanding the Southern "drawl," but as one cadet commented, Southerners did not come close to the Scots "with their colloquialisms and rolling r's."[106] John Evans, an "Arnold man," commented about his off-post experiences, "We were paid $10 a week, and when we got out of camp, which was not often, we had to scratch. Naturally, we had to hitch-hike, and we had to accept hospitality which was overwhelming, but we were very frustrated."[107]

On the other hand, some customs of the RAF cadets drew attention from the locals. The cadets tended to be younger and older, physically "less perfect," and less educated and of lower social status than their American counterparts. Comments were made on the British habit of bathing less frequently than Americans. Their habit of tea in the afternoon was interesting. Over time, however, the local community adjusted to the British presence. There were friendly and even "amorous reactions" between the local females and the RAF cadets, leading to "a fairly large number of marriages between RAF cadets and American girls."[108] For example, on the afternoon of March 6, 1942, Richard Ian Trotter from Newcastle-on-Tyme, England, who had graduated that morning from the advanced single-engine school at Craig Field in nearby Selma, Alabama, married Mary Walker Wells in a "ceremony of impressive simplicity" at St. John's Episcopal Church in Montgomery. John Clark of Surrey, England, assigned to the Gunter basic school as an instructor pilot, married Jean Harmon of Wetumpka, a small town about 28 miles northeast of Maxwell Field, at Southside (now Cloverdale) Baptist

3.12. Recently graduated Pilot Officer John Clark with Jean Harmon of Wetumpka, after their wedding in September 1942. When the RAF terminated flight training on Feb. 28, 1943, he returned to Britain as an RAF flight instructor, and his bride followed him six months later as an employee of the American Red Cross. See Appendix 9 for a short biography of Pilot Officer Clark.

Church in Montgomery in September 1942 after a seven-month courtship.[109]

As Montgomery was the site of the SEACTC headquarters and two AAF flight schools with RAF cadets, Maxwell Field received numerous official American and British officials as well as a number of contemporary personalities. GC Carnegie, the director of British Training in the United States, was a frequent visitor, often accompanied by members of the RAF Delegation and the British Air Commission in Washington. In October 1941, Air Marshal A. G. T. Garrod, the Air Member for Training, RAF, concerned about the high percentage of eliminees from AAF flight training, came to Canada and the United States. He, along with other senior RAF officers, talked to numbers of RAF and Royal Canadian Air Force members in Canada, met with RAF and AAF officials in Washington, and visited a number of schools in the Southeast. On January 9, 1942, Lieutenant General Arnold and Air Marshal Sir Charles Portal, the RAF chief of air staff, along with AAF Second Lieutenant James Stewart (perhaps better known as movie star Jimmy Stewart), visited Maxwell and Gunter fields. In 1943, Lord Halifax, British ambassador to the United States, and his wife; General George Marshall, US Army Chief of Staff; Sir Anthony Eden, British Foreign Secretary; and Field Marshal Sir John Dill, chief of the British foreign mission in the United States, visited Montgomery, just before graduation of the last RAF flight class on February 26, 1943. On April 4, 1943, President Roosevelt, Alabama Governor Chauncey Sparks, and Major General Barton K. Yount, commander of the Air Training Command, visited Maxwell Field. A number of other actors and actresses, including Al Jolson, also visited Maxwell and Gunter during the RAF training period.[110]

By mid-1942, the production of RAF pilots and other flight crew from the BCATP had reached a point where the RAF could terminate its training program in SEACTC schools and gradually curtail the training in the contract schools.[111] By the end of 1942, the BCATP schools had produced 69,115 aircrew, including 35,958 pilots.[112] In late June, the British Air Commission notified the office of Chief of the Air Corps (OCAC) that it no longer needed the AAF to train RAF crewmen and would be terminating the training and Class 43-B, starting its primary school on September 8, 1942, would be the last class of RAF students trained in SEACTC schools. The

3.13. RAF flight cadets pass in review, probably as part of the graduation ceremony from the basic school at Gunter Field, sometime in 1942.

British Air Commission notification to the Army Air Corps also terminated the training of RAF students at the Lancaster, California, and Sweetwater, Texas, contract schools, effective August 7, 1942, and the Pan American Airways Navigation School, Coral Gables, Florida, on August 28, 1942. The OCAC also notified the AAF Flying Training Command that it should fill those training slots with American cadets.[113]

Class 43-B began with 502 RAF students; on February 26, 1943, 342 (68.8 percent) of them graduated from the advanced school at Maxwell Field, ending the AAF's training of RAF flying cadets in SEACTC schools, including those at Maxwell and Gunter fields. The training of RAF flight cadets in the 35-week AAF program accomplished what the RAF could not do in Britain in summer 1941.[114] A day before Class 43-B's graduation ceremony, GC Hogan, director of training for the RAF in the United States, presented to Brigadier General William W. Welsh, the SEACTC chief of staff, the British flag and the RAF ensign that the AAF cadet detachment had presented to the RAF cadet detachment in October 1941 as a sign of friendship between the two countries. Welsh commented,

> The training in the Center marked a substantial contribution to

> the RAF when our need was greatest. We are grateful for the wonderful instruction our young men received. But we are also grateful for their presence here as a contribution toward lasting friendly relations of our two countries in time to come. These two flags were carried side by side with the American colors, as our British and American cadets marched together on parade.

General Welsh accepted the flags ". . . with a feeling of pleasure and sorrow as a token of the thousands of fine young British men who came here to receive their flying training and then departed. . . . These flags will become a part of the treasured archives of our training center."[115]

In total, 7,862 RAF cadets had entered 13 classes that spanned a 20-month period from June 1941 to February 1942 and saw 4,360 cadets (55.5 percent) complete the full program, with 78 killed in training accidents.[116] In July 1943, RAF Air Chief Marshal Sir Charles F. A. Portal wrote, "We are all most grateful to you for having allotted this training capacity to us. It has been of the utmost value at a time when our home facilities were restricted and the results in trained personnel and improved contact between our two Services have been of the highest order." Portal asked to have his comments forwarded to the staffs of the flight training schools.[117] AVM William Foster MacNeece Foster, deputy chief of the RAF Delegation, thanked the AAF Flying Training Command for the "splendid" training that the RAF students received in the United States since June 1941.[118] The official history of the Eastern Flying Training Command noted that, although the RAF and SEACTC had some initial start-up problems, overall, the Command believed that "[f]riendly cooperation with the British generally prevailed" and "the RAF program was well handled from its inception throughout."[119]

4

Free French Air Force Flight Training in Central Alabama

The defeat of France in June 1940 created a very complicated situation for Britain and especially the neutral United States. On June 22, 1940, representatives of the French government surrendered and signed an armistice with Germany. By the terms of the armistice, German forces occupied northern France and the Bay of Biscay coast to the Spanish border, the northern and western half of France. The World War I hero, Marshal Henri Petain, organized a government in the remaining southern half of France with its capital at the French spa of Vichy. To protect what was left of French sovereignty, this government strictly abided by the armistice terms and eventually collaborated with German authorities in the occupied zone. Further complicating this situation, President Roosevelt established full diplomatic relations with Vichy France and appointed Admiral William D. Leahy as U.S. ambassador.

Four days before the signing of the Franco-German armistice, Brigadier General Charles de Gaulle and a number of followers had escaped to Britain and issued a call for Frenchmen to rally to him alongside Britain. In early August, Prime Minister Churchill recognized de Gaulle as the head of all free Frenchmen and signed a formal agreement, the Charter of Free France, with de Gaulle. On September 24, 1941, the Free France organization established a French National Committee with de Gaulle as president.

Between August 1940 and November 1942, de Gaulle and his National Committee had worked to create Free French military forces from French manpower in Britain and the colonies which had declared their allegiance to de Gaulle. Several times in the first six months of 1942, he had asked American officials in London to allocate Lend-Lease equipment directly to his forces, but General Marshall, US Chief of Staff of the Army, believed that the Free French forces had limited striking power. By October 1942, de Gaulle had 35,000 men under arms, trained by the British and equipped with Lend-Lease equipment allocated to the British and not directly to the Free French. By the end of 1942, the Free French army consisted of two army divisions which had fought with the British Eighth Army in North Africa; the French "Navy" consisted of two navies, the Free French Navy which had operated under de Gaulle since fall 1940 and the North African navy which had survived the British attack in July 1940; and the Free French air force had five squadrons (two in the Middle East, two in Britain, and one in the Soviet Union). Additionally, by the end of 1942, French Equatorial Africa, the Cameroons, Syria, Madagascar, Djibouti, and Reunion broke with the Vichy government and declared their allegiance to de Gaulle.[120]

By mid-1942, the Combined Chiefs of Staff (CCS) had begun planning Operation Torch, the invasion of French Morocco and Algeria, scheduled for November under the overall command of General Dwight D. Eisenhower. Given the known animosity of Vichy French forces in North Africa for the British, the result of the Royal Navy's attack on the French Navy at Mers el-Kebir on July 3, 1940, two-thirds of the invasion forces were Americans (83,000) and one-third British (26,000). American authorities were reluctant to divulge too much information to de Gaulle and his headquarters because of suspected security leaks and excluded de Gaulle from the planning and execution of the invasion. Instead, American planners had opened negotiations with General Henri Giraud whom they believed would "most likely" lead French North Africa and West Africa to the Allied side. On October 23, Lieutenant General Mark Clark secretly landed in Algeria to present assurances of direct American aid for the French North African forces to a group of French military leaders in exchange for recognizing Giraud as leader of the French forces in North Africa. The invasion occurred on November 8–7, and on November 10, Admiral François Darlan, the commander-in-chief of Vichy French military forces, ordered French forces in North Africa to surrender. Unfortunately (or fortunately, depending on one's point of view), Darlan was assassinated on December 24, 1942, clearing the way for Giraud to become the senior military leader of French forces in North Africa.[121]

Allied military advances in North Africa in 1943 paved the way for

Lend-Lease military aid to the French forces in North Africa and the start of French armament in earnest. By May, the Americans and British forces, advancing toward each other from Algeria and from the Libyan-Egyptian border, respectively, progressively squeezed the Axis forces into an ever smaller area toward the port of Tunis and forced their surrender on May 12, 1943. As a result, the CCS and U.S. forces in North Africa could now focus more on training and equipping additional French forces, including a French air force, for upcoming military operations in the Mediterranean. In November 1942, General Arnold urged caution to Major General Carl A. Spaatz, commander of Twelfth Air Force in North Africa, on utilizing French pilots, considering the determination of loyalty, need for security, language differences, ability to absorb training, and familiarization with American equipment. Spaatz reported in December 1942 that the French had about 14,000 men, including 1,500 pilots and associated crews, sufficient to operate a force of 300 fighters, 400 medium bombers, and 100 transports. Most had seen combat in the 1940 French campaign, had served in North Africa, and/or had fled to North Africa from France. Many were anxious "to get into the fight" against the Axis.[122]

In January 1943, the French Air Force in North Africa consisted of 20 fighter, bomber, reconnaissance, and transport squadrons throughout North Africa and ten more in French West Africa but with only several hundred combat worthy aircraft out of some 700. As the senior U.S. and Free French leaders discussed plans for French rearmament, US and Free French officials in North Africa took a more realistic approach to developing a Free French air force in North Africa. In the early months of 1943, substantial numbers of pilots, crews and mechanics had attended an Allied air training center in Africa. In March 1943, Spaatz announced that his XII Air Force Training and Replacement Command would oversee the training of French combat crews, assigned to French air units attached to the North African Air Forces (NAAF).[123]

In addition, the French opened several schools in Morocco, the French Air School, a school for practical training of flying personnel in Marrakech, a school for advanced training in Casbah-Tadla, and a school for firing and specialists' training in Agadir, with largely American and British materiel and personnel assistance for those French air crew with a certain level of flying experience.[124]

The Free French needed both flight training for replacement air crews and contemporary aircraft to replace obsolete models. In February 1943, Lieutenant General Emile Bethouart, then Chief of the French Military Mission in the United States, submitted the first Free French air armament plan. After studying the proposal, Arnold recommended that the CCS immediately determine the extent, composition, utilization and training, and equipment for the North African Air Forces. Already, AAF units in Algeria and French Morocco were training French pilots and air crews. Spaatz wanted to send four successive contingents of 500 student pilots to the United States for flight training, but, because of the AAF's training load in the United States, the AAF in February 1943 recommended a more modest program of 100 French trainees to start on June 28, followed by another class of 100 in July, and 50 trainees in following classes.[125] Also, General Eisenhower had recommended to the War Department to provide aircraft to equip at least one light bomber group and one fighter group. In response, the War Department notified Eisenhower and Spaatz of plans to train French pilots in the United States, starting in June, and the initial shipment of P-39 fighters, 67 A-35 dive bombers, and 60 C-78 transports in March or April although the CCS had not yet made any decision on their requests.[126]

By the summer of 1943, when the United States agreed to train Free French pilot cadets, France itself was still occupied by the Germans; the French National Committee headed by General Charles de Gaulle directly represented only a minority of the French people; and US and British forces had recently freed its North African colonies of Axis troops. The end of the flight training program for the RAF in February 1943 opened up training capacity and facilities for the French at Gunter Field and other AAF flight training schools. The training of French flight cadets developed much differently than had that of the RAF cadets. In 1941, when the United States government agreed to train British flyers, the US was busy expanding its own military forces at the same time. Thus, for several months both AAC and RAF cadets had to deal with shortages of facilities, instructors, and training aircraft. By 1943, these issues were resolved or ameliorated. But the training of Free French flight crewmen in central Alabama from summer 1943 to October 1945 experienced many of the other problems that had developed in the RAF training, such as the AAF honor and disciplinary systems, as well as more difficult language problems.

In April 1943, SEACTC received official notification that Free French cadets would soon arrive at designated schools for flight and air crew training. The initial message stated 20 students would receive training in flexible gunnery at the Tyndall Field, Florida to the proficiency level of American gunners in a longer course because of the language difficulties. The notification also recommended that the French officer in charge should speak some English and that the school provide two French-speaking instructors and an interpreter. A second letter in late April informed SEACTC that it would train both French pilots and flexible gunners. The final plan called for training a total of 500 pilots. A hundred French cadets would enter Class 44-A of primary training on July 2, 1943, a hundred more, a month later on July 30, and 50 more each month afterwards. In addition, 20 French students would enter the flexible gunnery course on April 19 and 20 additional students in each month afterwards. Following completion of advanced single-engine course, 50 percent of the graduates would enter a Replacement Training Unit (RTU) for fighter training and the other 50 percent would enter a dive bomber RTU.[127] The AAF requested funding to cover the entire cost of French flight training, including training costs, transportation within the United States, housing, medical care, individual clothing and equipment, and so on, through the Lend-Lease Act.[128]

Several weeks later, SEACTC informed the Flying Training Command of how it intended to handle the French flight training. Since the French cadets would be arriving in the United States in small groups and at unspecified intervals, due to the uncertainties of transportation from North Africa, SEACTC planned to pool them at Craig Field, Alabama, until sufficient cadets had arrived to begin a class with a full complement. The Alabama Institute of Aeronautics, Tuscaloosa, Alabama, would provide primary training for the French. They would then receive basic training at Gunter Field, followed by advanced single-engine training at Craig Field or advanced

4.1. Layout for a naval vessel recognition test for the French cadets in the ground school portion of the primary flight school at Craig Field, Alabama, in late 1943.

twin-engine training at Turner Field, Georgia.[129] Those French cadets who completed advanced training would receive their wings, but only the *Eleves Aspirants Reserve* who were made *aspirants* and some *aspirants* who were made *lieutenants* would receive officer commissions.[130] From advanced training, most of the French trainees, unlike the RAF cadets, went on to some type of transition training in specific types of US aircraft before they returned to North Africa, Britain, and, after June 1944, France.

SEACTC also provided initial plans for the administration of the French students. One captain and two lieutenants would command the first detachment of 100 students, reporting to Craig Field. The detachment would be divided into ten groups of ten students with each group having several of them able to speak English.[131] When the first 100 students appeared at Tuscaloosa on July 2, there were also two officers with the group who stated that they were also students. The French Military Mission in the United States dispatched Captain Phillipe Morize to Tuscaloosa to command the French student pilots at the flying school. He would also complete the primary training there and accompany the graduates through the completion of their flight training. However, SEACTC headquarters preferred that one or two French officers be permanently assigned to each SEACTC school with French students to learn the administrative rules, feeding and living rules and other routine issues and explain these rules to the French students upon their arrival on station.[132]

From the start, SEACTC officials recognized that the language difference would require solutions that they had not encountered with the RAF cadets. Mr. Oliver Parks, director of the contract flying school at Tuscaloosa, foresaw that his primary school would need additional French-speaking instructors to achieve a ratio of one French-speaking instructor for every three French students instead of one instructor per every six American students because of a probable lack of fluency in French technical terms and phraseology, related to aircraft and flying. The difficulty in determining the expected elimination rate of the prospective French students added to the difficulty in determining the exact number of French-speaking instructors needed for the training. Additionally, the need to obtain French-speaking instructors from across the country, transporting them to Alabama, for the

4.2. French-speaking flight instructor Lt. Norbet J. Cain, Syracuse, NY, explains the instruments of the Vultee BT-13 trainer, used at the Gunter Field Basic Flight School, to Free French flight cadets, late 1943.

period of the training added to the cost of training. A survey of SEACTC revealed that the command had five French-speaking advanced single-engine instructors, eight basic instructors, and 38 civilian primary instructors. The same survey revealed the only about half of the instructors would relocate to Tuscaloosa. Finally, SEACTC estimated there should be one instructor for every three primary school students and one instructor for every four basic and advanced students.[133]

During the summer, the AAF and SEACTC modified these original plans for Free French flight training in the United States. SEACTC changed the percentages for the French basic school students graduates to 60 percent to attend advanced single-engine training and to 40 percent of graduates to attend advanced twin-engine training. Of those graduating from advanced single-engine training, five-sixths would receive fighter training and one-sixth would receive dive bomber training. Of those graduating from twin-engine school, 50 percent would go to "RTU" training and the other 50 percent

4.3. US flight instructor (l) with a model aircraft and French flight cadet discuss recent flight lesson with use of a French-speaking translator at Van de Graaf Field, Tuscaloosa, Alabama, primary school.

would transition to transports. French cadets eliminated from flying training would receive flexible gunnery, armament, radio, or bombardier training. By mid-October, the French and US officials had changed the number of pilots to enter primary training and, on November 5, 1943, increased the number from 100 to 132 each month. They hoped that at least 79 would graduate, and the eliminees, augmented with additional French enlisted men, would enter training as bombardier-navigators, navigators, radio operators, and flexible gunners.[134]

The first group of French flight cadets, numbering 104, left North Africa, where most of them had served in ground crews or paratrooper units, in late May by ship and disembarked at an Atlantic port, took a train to Selma, Alabama, and arrived at Craig Field in June 1943 for "in processing." All in the early twenties, they were already veterans of the war, either in France or in North Africa, and some had gone through harrowing experiences to get to North Africa.[135] They entered primary training at the AAF contract school at Tuscaloosa on July 5, 1943. In preparation for the training, the contractor had provided eight interpreters, naturalized French-Americans or Americans who had been educated in France, who spoke, read, and translated French. The ground school held French classes daily for both military and civilians to familiarize them with the sound of French and to master a few words, and the assigned interpreters translated training manuals, examination forms, and instructions. The interpreters attended the classes for the French students and translated the instructor's words, sentence by sentence, into French. Initially, the long classroom hours and the "lack of understanding of the French mind" caused some difficulties between the instructors and the French students, problems which healed over time and led to new teaching methods. Interpreters also accompanied the students to the flight line and explained the day's maneuvers to the students. Because of the language difference, most flight instructors and students could not directly communicate with each other. During the flying lesson, the instructor pilot and his student used hand signals to communicate during their flying lesson, and the instructor had to wait until landing and the appearance of an interpreter to review the flying lessons with the student. Overall, the French students required about seven hours of instruction as opposed to an average of five hours for American students.[136]

As the first class of French cadets started their primary training at Craig Field in early July, the basic school at Gunter began preparations to receive the graduates of the primary school. The first major problem was obtaining sufficient numbers of qualified French-speaking instructors to meet the specified one instructor to four student ratio. The first group of French-speaking officers arrived in mid-June who began classes in French to improve their oral French and learn technical terms and began to translate the English-language manuals and tests into French.[137]

A month later Gunter was notified that it would receive 19 French-speaking officers as instructors from other AAF schools and that it would, in turn, transfer an equal number out as replacements. Four of these instructors would teach ground school subjects to the French cadets, and two of these would also teach Americans. However, the school discovered that many of

4.4. French flight cadets salute Maj Gen Marie E. Bethouard, Chief of the French Mission in the United States, during a visit to Gunter Field, Alabama, on Oct 15, 1943.

these instructors were not qualified, and the school commander requested 11 additional instructors and would be required to transfer out only seven instructors. Headquarters EFTC acknowledged the request but believed that the French cadets would have reached a level of English proficiency by the time that they reached Gunter that it would not need to furnish all of the requested French-speaking instructors. In late July, the school commandant notified SEACTC headquarters that it was still short 45 instructors, administrators, and control tower operators.[138] As late as November 1943, EFTC schools apparently were still receiving French-speaking instructors with insufficient command of French to conduct the flight training program in French and recommended to Headquarters, AAF Flying Training Command that it should interview prospective instructors before assigning them to a school with French cadets.[139]

The first class of French primary school graduates, 75 in number, ranging in ages 18 to 28, including Captain Morize, arrived at the basic school at Gunter Field on September 1 with the French flag flying in the breeze to begin the basic training school phase. The French cadets wore US Army uniforms but with a shoulder patch that consisted of a red, white, blue circle

(French tricolor) with "France" above it and a garrison cap topped in white, instead of khaki, and a flight cap bordered with red, white, and blue piping.[140] The AAF selected Lieutenant Leon Basier as the French liaison officer. Born in Lille, France, he had come from France as an orphan after World War I, had received a PHD at Catholic University, Washington, DC, returned to France in 1939, came back to the United States, enlisted in the US Army in September 1942, and received a direct commission in June 1943. He compiled a French-English aviation dictionary for use at Gunter Field, acquired the French-speaking instructors for the basic school and taught them the technical aspects of the French language. SEACTC brought in seven French-speaking officers to teach the ground school in the same subjects as the Americans but translated into French and 20 French-speaking enlisted men to take care of the French cadets' administrative issues.[141]

During this class, the French cadets participated in a parade in Montgomery during which they displayed the French flag to the cheers and applause of the crowd. The French cadets completed their training at Gunter "with flying colors" and "a zealous enthusiasm in every phase of training."[142]

Out of 75 students entering Class 44-A, 72 graduated, one was eliminated, and two were held over to the next class (graduation rate of 98.6 percent). The next class, 44-B, had 65 graduates out of 68 students (91.5 percent), one student killed, and five eliminated. Of the graduates of Class 44-A, 48 went to advanced single-engine training, and the remaining 23 attended (and graduated from) the advanced twin-engine school at Turner Field.[143]

The French authorities in North Africa had selected the students who had rated the highest on tests given in North Africa for these first groups for training in the United States. Thus, apparently seeing the training as a reward for performance in North Africa, they were willing to study hard and were eager to learn how to fly. They appeared to want to make most of their training opportunity as some had waited as long as four years for the chance to strike back at the Germans and hopefully liberate their homeland. They tended to be more attentive in class, demonstrated excellent discipline, and appeared to be more interested in theory than their American counterparts. As a result, they appreciated the ground school portion of the training more than the Americans who thought that only flying was important. They were

4.5. Col A. J. P. Le Bel, Chief of Staff of the French Military Commission; Maj A. de Ponton d'Amecourt, Chief of the French Air Training Centers; Lt Gen Paul Entienne Beynet, Chief of the French Military Mission; 1Lt François Thore, Commandant of the French cadets at Gunter Field; and 2Lt Louis Bouvard, Lt Thore's assistant, left to right, during a visit at Gunter Field on Jan 12, 1944.

generally more careful with their aircraft than the RAF cadets who had preceded them and, as a result, had an accident rate comparable to that of the Americans. The French cadets also worked harder than the Americans as demonstrated by generally higher grades for the academic phases of the training program with the exception of radio code, probably because the French cadets had no real pre-flight training while the Americans did during which they received 48 to 50 hours of radio code training.[144]

Perhaps the most significant problem after the language differences was the status of the French cadets and their perceived benefits from that status. The vast majority of the French cadets were enlisted men, but a few were "*aspirants*," essentially an officer candidate without a warrant or commission. Many of these *aspirants* desired and, in many cases, insisted on the privileges accorded to officers, such as open post, late curfews, and membership in the post officer's club. They did not like preparing their rooms for inspections and resented taking orders from American officers. The American officers at the AAF schools did not understand the strict sense of rank held by the *aspirants*, many of whom had been officers in the French Air Force before and during the early years of the war. The Americans either treated them as enlisted men, commingled the officers with French enlisted personnel, or had American enlisted men giving orders to the *aspirants*. Because of these misunderstanding, a number of French students were apparently eliminated from the flight schools.[145]

By early 1944, the problems that the French cadets were having in adapting to the AAF disciplinary system at all of the AAF schools resulted in the EFTC calling a conference to discuss these problems and develop solutions. The conference confirmed that the French commanding officers at each AAF base fell under the command authority of the Chief of the French Student Corps in the United States but this authority was delegated to the post commanders of each AAF installation with French cadets. Secondly, the French trainees were not cadets but either officer students, warrant and noncommissioned officer students, flight officer candidates (*Eleves Aspirants Reserve* [EAR]) and enlisted men (grades below sergeant). The base commander would transmit all administrative orders to the post French commanding officer. While instructors could still administer administrative discipline, related to training issues, only the French commandant could administer punishment, according to French Army regulations, not AAF regulations, upon the recommendation by the post commander. In addition, the conference recommended that a French officer be appointed to represent the Chief of the French Student Corps and assigned to EFTC headquarters at Maxwell Field. Lieutenant Ferreol Bols, the commandant of cadets at Turner Field, was assigned to Maxwell Field in this new position on March 12. The implementation of these decisions not only significantly reduced the number of serious problems with the French cadets at the EFTC schools but also reduced the time to resolve problems that would have gone to Washington, DC, for resolution.[146]

With the start of 1944, there were still significant problems with the training of French cadets at the EFTC schools. Since the language difference was so significant, in April 1944, the EFTC began teaching basic English to the French cadets in the pool in Craig Field, using two French-speaking ground school instructors who spoke French fluently but without a text, which was expected in May. Also, there was no preflight instruction program which placed the French cadets behind the American cadets in radio code proficiency so the French trainees began receiving ten hours of code instruction by the end of April. Also, the basic school at Gunter transferred the administration of the French detachment to French officers and introduced additional distinctions between the French and American groups of trainees. For example, the school provided diagrams of all traffic patterns, auxiliary fields, cross-country flights, and night-flying; and more written materials, regulations, and questionnaires in French for the French cadets than for the Americans. In addition to language differences, the physical standard of the French students, specifically vision, apparently contributed to the relatively high accident rate among the French students at Gunter. Regular lectures on safety, vigorous corrective action after most minor accidents, and prohibitions on night-flying contributed to lowering the accident rate among the French cadets. Eventually, the EFTC decided to reassign French cadets who could not meet American physical requirements to other duties. Additionally, the school tightened discipline on the flight line by limiting the interaction between the French trainees and their instructors to actual training periods. Finally, the EFTC transferred eight French-speaking American cadets in Classes 44-B and 44-C at Columbus and Moody Fields to Gunter as instructors. These measures resulted in reducing the numbers of accidents.[147]

The periodic EFTC and the Gunter Field histories generally provided few details about the specific causes of aircraft accidents at the basic school during the period of the Free French training program, much less the causes of accidents that involved French cadets. In December 1944, the commander of the basic school apparently directed the director of training and operations to conduct a detail study as the Gunter Field history for the period November 1 through December 1944 had such a study of training accidents for 1944. The study revealed that Gunter Field experienced a total of 139 accidents, 75 during the period January through June 1944 of which

43 were solo and 32 were dual (student and instructor), and 64 accidents during the period July through December of which 32 were solo and 32 were dual. Tables 4.1 and 4.2 show the statistics by cause of accident and by nationality.

Table 4.1.
Causes of Gunter Field Accidents for 1944

	Jan–June Solo/Dual	Jan–June % of Total	Jul–Dec Solo/Dual	Jul–Dec % of Total
Poor Landings	25/20	60.0	1/15	40.83
Taxiing	7/3	13.33	8/7	23.62
Nose-Ups	4/1	6.67	1/3	6.45
Ground Loops	1/0	1.33	4/2	8.20
Stalls & Spirals after Take-Off	3/2	6.67	1/0	1.65
Low Flying	2/2	5.33	0/0	0.00
Mid-Air Collisions	2/0	2.67	3/2	8.09
Other Causes	1/2	4.00	4/3	11.08
Totals	75 (43/32)	100.00	64 (32/32)	100.00

Table 4.2.
Nationality Statistics for Gunter Field Accidents for 1944

	Jan–June	Jan–June % of Total	Jul–Dec	Jul–Dec % of Total
American	55	74.65	50	78.12
French	20	25.35	14	21.85
Total	75	100.00	64	100.00

Source: Letter, Lt Col Carver T. Bussey, Gunter Basic School Director of Training and Operations, to all supervisory personnel, "Accident Analysis 1944," Jan 1, 1945, Appendix B, *Gunter Field History, 1 Nov 1944 to 31 Dec1944*, AFHRA Call # 283.81–6, Doc #00172969.

While one cannot be conclusive without additional data, it appears that over the course of the French training program, June 1943 to October 1945, the American cadets as a percentage of the total accidents were involved in about 75 percent of the accidents and the French cadets about 25 percent. By looking at Table 9, the American to French ratio for each

4.6. Members of the Free French Military Mission to the US inspect French cadets at the basic school, Craig Field, Alabama, late 1943.

class was roughly the same percentage, meaning that, while the percentages for Americans and French cadets causing accidents for specific classes may be different with French cadets involved in slightly higher percentage of accidents, overall the respective percentages were probably roughly the same for 1943 and 1945 as they were for 1944.[148]

During the first half of 1944, the Gunter basic school added two evening programs to the French training to help the French cadets in their training. The basic school provided classes in English, twice a week, which the cadets appreciated. The French cadets also received additional radio code training at night to bring them up to the same proficiency level as the American cadets who received 50 hours of code training in their preflight school which did not yet exist for the French cadets. However, the additional evening training, which lengthened their training day to 16 hours, "wreaked havoc with the French desire to relax and reflect." After a while, the school terminated the English and code classes.[149]

With a year of experience in training French cadets, the AAF and the

French liaison officer at Headquarters, Eastern Flying Training Command (HQ EFTC) in July 1944 decided to develop a formal pre-flight instruction program for newly arrived cadets at Craig Field to remedy noted training deficiencies of the French cadets and a pre-flight school at Maxwell Field for eliminees from various training schools. Previously, the French cadets, waiting at Craig Field to enter a specified class for pilot or gunnery training, received some training in radio code, a course in basic English, and various training films but no formal program of instruction. At the same time, schools training French cadets returned eliminees from their programs to Craig Field where they mixed with newly arrived French cadets, apparently causing morale problems by spreading "ill founded gossip and rumors" to the pool cadets. As a result, the representatives of the HQ EFTC, the Craig Field director of training, the director of the ground school, the French commandant at Craig Field, and the French liaison officer, Captain Robert L. Lamaison, from HQ EFTC developed the following programs:[150]

TABLE 4.3.
INSTRUCTION PROGRAMS FOR FRENCH PRE-FLIGHT SCHOOL AND ELIMINEES

Pre-Flight Program Instruction		*Program for Eliminees*	
Subject	**Hours**	**Subject**	**Hours**
Basic English	40	Basic English	2 hrs, 5 days per week
Small Arms Firing	5	Small Arms Firing	Until Qualification
Mathematics	40	Radio Code	1 hr, 5 days per week
Radio Code	25	Recognition	3 hrs per week
Principle of Flight	15	Physical Training	1 hr, 5 days per week
Aircraft Recognition	20	Military Drill	
Naval Recognition	10		
Orientation Lectures	5		
Physical Training	40		
High Altitude Training	Weekends		
Military Drill			

Source: *AAF EFTC History, 1 Jul 1944 – 31 Aug 1944*, vol. 1, p. 292, AFHRA Call # 220.01 1 Jul 1944 – 31 Aug 1944 v. 1, Doc # 001461521.

As in the past, a significant problem was obtaining sufficient numbers of French-speaking instructors for the pre-flight school at Craig, a problem solved partially by obtaining instructors from the pre-flight school at Maxwell.[151]

On September 8, 1944, about 100 French eliminees from flight training at Orangeburg, SC; Gunter Field, and Craig Field and about 47 student gunners from Algiers, North Africa, at the request of Lieutenant P. Carpentier, a French liaison officer at Maxwell, arrived at Maxwell from Craig Field for the preflight school. The installation arranged messing and housing for the students after which it held a "short but impressive" ceremony to officially raise the French flag. The training program officially began on September 11. The cadets' training for the first week included gunnery, military drill, range, and physical training. On September 20, the Plans and Training Office with discussion with the two French liaison officers at Maxwell, Lieutenants Bole and Carpentier submitted a training program, consisting of weapons, chemical warfare, code, aircraft and naval recognition, physical training, and close order drill. In October, the training office deleted chemical warfare and added a 14-hour block on the principles of electricity. Navigators and bombardiers also received training in maps and charts and the E-6B computer, used for flight planning. On December 11, the program added beginning and advanced English courses.[152] By December 1944, after receiving ground instruction at the Maxwell Pre-Flight School, about 75 French students, classified as gunners, navigators, bombardiers, mechanics, and radio operators, left Maxwell monthly for various AAF training schools. The EFTC discontinued the preflight school at Maxwell on December 1, 1944, and now sent American flight training eliminees to Keesler Field, Mississippi.[153]

The French cadets, eliminated from flying training, however, continued to come to Maxwell to a French-only "preflight school," designated as the French Reclassification Center, with a staff of two officers and six enlisted men. The number of French cadets varied during January and February 1945, averaging about 160.[154] In March 1945, the center received 87 eliminees and lost 64 reclassified students. In April, 84 arrived and 65 reclassified cadets left. During these two months the total at any given time ranged

Table 4.4.

FFAF Statistics for the Gunter Basic School

Class	Start Date	Fr Entered	Amer Entered	Fr Held Over	Amer Held Over	Fr Eliminated	Amer Eliminated	Fr Killed	Amer Killed	Fr Graduated	Amer Graduated	% Fr Grad.	% Amer Grad.
44-A	9/1/43	75	290	2	14	1	9	0	0	72	267	98.6	96.7
44-B	10/3/43	69	294	0	18	5	12	1	0	65	264	91.5	95.7
44-C	11/5/43	76	335	7	68	5	19	0	0	64	248	98.7	92.9
44-D	12/ 7/43	73	300	5	24	5	18	0	0	70	258	93.3	93.5
44-E	1/17/44	93	291	8	15	14	22	0	0	77	254	84.6	92.0
44-F	3/27/44	90	292	6	2	12	36	0	0	80	254	87.0	87.6
44-G	4/22/44	104	252	11	7	9	22	0	0	89	223	90.8	91.0
44-H	5/26/44	11	287	8	5	30	29	0	0	0	253	0	89.7
44-I	5/26/44	62	217	4	7	13	32	0	0	45	177	77.6	84.3
44-I-1		87		2		29		0	0	64		68.8	
44-J	6/30/44	61	294	6	7	14	38	0	0	41	249	74.5	86.8
44-K	8/7/44	63	334	2	9	16	49	0	0	45	286	73.8	85.4
45-A	9/11/44	53	298	3	9	21	64	0	1	29	224	58.0	77.5
45-B	11/23/44	110	248	4	10	33	20	1	0	66	217	66.0	91.6
45-C	12/31/44	92	91	5	7	17	15	0	0	70	69	80.5	82.1
45-D	2/7/45	112	115	9	7	20	17	0	0	82	83	79.6	79.0
45-E	3/18/45	99	87	7	5	17	12	0	0	75	70	81.5	85.4
45-F	6/30/45	107	102	6	4	27	24	1	0	73	74	72.3	75.5
45-G	5/26/45	109	122	6	10	9	26	1	0	93	86	90.3	76.8
45-H	6/30/45	124	650	7	6	17	14	0	0	100	45	85.5	76.3
Totals		1,670	4,891	108	234	314	478	4	1	1,300	3,601	77.8	73.6

Source: AAFFTC Pilot Class Histories, Total Training, Classes 44-A through 45-H, *AAF EFTC History, 1 Sep 1945 – 15 Dec 1945*, Vol. 5, Appendix, AFHRA Call # 222.01 1 Sep 1945 – 15 Dec 1945 v. 5, Doc # 00146165. There were no in Americans in Class 44-I-1 Field.

from 200 to 243 with an average of 200. Between March 30 and April, 16, the French eliminees received demonstrations in the B-29 mock-ups, used the B-29 transition training at Maxwell.[155] On June 6, 1945, the EFTC moved the center to Craig Field and assigned the AAF personnel to other positions at Maxwell Field.[156]

However, growing dissatisfaction among the French cadets with the facilities at the pre-flight school at Craig Field led Captain Robert Lamaison, representative of the French Air Forces Students in the United States and assigned to Headquarters, EFTC at Maxwell Field, in December 1944 to recommend that the EFTC move the Craig pre-flight school to Maxwell along with the Reclassification Center. He believed that Montgomery would offer the cadets more opportunities for recreation and social life, allowing the cadets from North Africa to learn more about American life and culture than the smaller city of Selma. Additionally, many of the cadets were relatively uneducated and knew little about American life, history, and culture but found themselves plunged into a strenuous flight training program within five days of arrival at Craig. [**See Table 4.4.**] Thus, the cadets believed they needed more time to acclimate to the United States, which meant increasing the length of the pre-flight school. Only the French Air Mission in Washington had the authority to increase the length of the school. However, instead of moving the school to Maxwell, the EFTC made improvements at the school's facilities at Craig.[157]

By summer 1944, the training of French cadets was operating more or less smoothly. In June the EFTC also decided, beginning with Class 44-J, to send 60 percent of the French graduates from the primary school at Craig to single engine basic training at Gunter and the remaining 40 percent to two engine training at Shaw Field, South Carolina. While the cadets at Gunter would complete the course in single-engine aircraft, the French cadets, selected to attend B-25 training in an advanced school, would receive training in the AT-10 twin-engine aircraft in the last week of their basic training at Shaw. However, by the end of September, the EFTC had reversed itself, and the Class 45-C graduates of the primary school, scheduled to enter the basic school on November 3, 1944, would be the last French cadets sent to Shaw Field as the EFTC wanted the French cadets, scheduled

4.7. Charles de Gaulle, leader of the Free French, meets Free French flight cadets at Gunter Field, circa 1944.

for B-25 type aircraft in the twin-engine advanced school, to have as much time as possible in an AT-6 trainer. This shift meant that Gunter would see an increase in the number of students from the primary school from 60 to 100 per class. To handle the increased student load, the EFTC transferred a number of the instructors from Shaw to Gunter.[158] Overall, the percentage of French eliminees increased over the course of the year (average of 10 percent to 25–35 percent), attributed partially to the EFTC September 1, 1944 directive which required the French cadets to attain the same level of proficiency as the Americans trainees and to the makeup of the later classes with French North Africans versus French cadets from France proper. [159]

The training of French cadets continued into 1945. The first two groups of French cadets were composed of specially selected men, but those for subsequent classes were French Army volunteers, often with no special qualifications.[160] By early 1945, some of the cadets had been members of the French Forces of the Interior, former French resistance members, who did not have the education level of previous flight cadets. While they generally made up for this deficiency through hard work, they also tended to be more temperamental as evidenced by their approach to the classroom and exams. The language and cultural differences continued to affect relations between the French cadets and their American instructors. Sometimes, the cadets felt that the instructors might misunderstood their reactions, attitudes, or gestures. The cadets felt that the instructors were too concerned about cadet eliminations, overused threats to eliminate cadets for training problems, and did not provide sufficient praise to increase the confidence of the cadets. Yet, many French students preferred the American training system and discipline as it tended to correct errors while the French system tended to punish students.[161] From Table 11, despite the cultural differences, language problems, and perceptions of ill-treatment by the French cadets from their American instructors, the elimination rates of the French cadets in the basic schools in 1944 generally paralleled those of the American cadets until the latter part of the year.[162]

Table 4.5.
Elimination Rates for EFTC Schools, June 1943–Nov. 1945

Phase of Training	American (%)	French (%)
Primary	29.2	30.2
Basic	20.1	22.4
Advanced	15.4	17.7
Overall	51.8	53.7

Source: Based on class statistics from the AAFFTC Pilot Class Histories, Total Training, Classes 44-A through 45-H, EFTC History, 1 Sep 1945 – 15 Dec 1945, Vol. 5, Appendix, AFHRA Call # 222.01,V. 5, Doc # 00146165.

The French cadets also had some opinions about American customs and culture. They generally liked American food but found it less tasty than theirs. They preferred classical music but could not understand why Americans liked American jazz, given the dislike of some Americans for blacks. The French cadets, interested in philosophy, history, and novels (serious reading) while Americans liked less serious reading, felt that the Gunter library was very good. Like most people far from home, they looked forward to mail from their families and worried about the length of time it took mail to reach them and their families. Because of the language differences, social activities for them in Montgomery were limited, but still many of them received invitations for dinner and entertainment in local Montgomery homes. Like the RAF cadets before them, the French cadets participated in military parades in Montgomery during which they displayed the French tricolor for the first time in a military parade since the fall of France in June 1940 and to the cheers of "Vive la France" from the Montgomerians. Many also spent their weekends in Birmingham, about 80 miles north of Montgomery, which had more interests for them and a larger number of French-speaking Americans.[163]

The primary trainer for the AAF basic schools, including the Gunter school, from its inception in 1941 until mid-1944 was the Vultee BT-13 Valiant. In June, the AAF began replacing the BT-13 with the AT-6 as the trainer for the basic school, and by early 1945, Gunter was about 65 percent complete. The Gunter Basic School completed the conversion to the higher performance trainer on February 14, 1945, and Class 45-D was the first class to use the AT-6 exclusively. The initial misgivings of the students in flying the advanced trainer in the basic school quickly dissipated, after acclimating themselves to the AT-6's more cluttered cockpit, and they had few problems in transitioning from the BT-13 to the higher performance AT-6. Experiments conducted in the latter part of 1943 indicated that the normal student experienced no problems in transitioning from the primary trainer to the advanced trainer. The school made some adjustments to the training curriculum to account for differences in performance characteristics between the two aircraft. Although the transition from the BT-13 to the AT-6 as the training aircraft for the basic class resulted in a slightly higher elimination rate among the French cadets, the French cadets generally did not find the transition that difficult.[164]

With the surrender of Japan on September 2, 1945, World War II came to an end. On August 6, the Americans and French students of Class 45-G transferred to Craig and Turner fields for advanced training, and Class 45-I, consisting of French students only, arrived the next day from the primary school at Orangeburg, SC. On September 10, Class 45-H, consisting of

112 French and 15 Americans, transferred to an advanced school, and the last class, 45-I, transferred to Spence Field, near Moultrie, Georgia, to complete their training. The departure of these two classes brought to a close almost five years of continuous flying training operations at Gunter and Maxwell Fields.[165]

By December 1944, the Free French Air Force consisted of 25 operational squadrons, flying predominantly American-made aircraft.[166] With the end of the war in Europe in May 1945 and no plans to use the French Air Force in the Pacific Theater, the War Department wanted to terminate the French training program in the United States. However, the French government "exceedingly" wanted to continue their flight training programs. On May 25, 1945, the War Department notified the chief of the French Air Mission in Washington that it would no longer accept new French students beyond the informal agreement to train specific monthly P-47 and B-26 crews but would complete the training of French students already in the United States. On November 1, 1945, Headquarters AAF announced that it would terminate all French training under Lend-Lease on November 30. After a French protest, President Harry Truman agreed to complete the training of French students in the United States up to March 1, 1946 if the French government paid for the training after November 30. Since the French government did not have the necessary funds, all French training came to an end on January 18, 1946.[167]

By that time, the AAF had trained 1,351 pilots, 253 bombardiers, 74 navigators, and 955 flexible gunners, undoubtedly "a substantial fraction of the existing French Air Force."[168] (See Appendix 7 for statistics of the FFAF cadets at the EFTC training schools, June 1943 through November 1945.) Generally, the relatively small FFAF received favorable comments, such as "small but damned good" and "second to none," from a number of Allied military leaders about its performance in the European-Mediterranean theater.[169] As for the French airmen who received the opportunity to train in the flying schools of the AAF during World War II from June 1943 to the end of 1945, they demonstrated "a drive, an intensity, with which they go at their flying or their ground school which is unusual." In the end, they "appreciated the opportunity to learn how to fly so that they could return to North Africa and later France and go after the German forces which conquered their lands, and separated them . . . from the people they love."[170]

5

Results of RAF and FFAF Training in Central Alabama

The US government terminated all training of international students except for those from China and a few Latin American countries under the Lend-Lease Act on March 2, 1946.[171] During World War II, the AAF trained over 21,000 aircrew members (pilots, navigators, bombardiers, gunners, flight engineers, and technicians) from 31 different Allied nations. By the far, the top three countries were Britain (over 12,300), France (over 4,100), and China (over 2,000) with the remaining 4,900 divided among 28 different countries). The 12,300 from the RAF and the FFAF represented 78.5 percent of the total trained during the war (59 percent and 19.5 percent, respectively). In comparison to the overall AAF training program, which trained over 1.5 million from all phases of flight and technical training, this number represents s mere 2.0 percent of the total. However, for the RAF and the FFAF, the training of RAF and FFAF aircrew members in the AAF flight schools, especially during the phases at Gunter and Maxwell Fields, had, in a real although relative sense, a significant impact on the combat operations of the RAF and the FFAF during the war and, more importantly, inter-Allied relationships. In addition, the flight training of these Allied aircrew members foreshadowed the postwar flight training by the post-1947 US Air Force of pilots and other aircrew members from allied nations, particularly those of the North Atlantic Treaty Organization (NATO) after the start of the Cold War and the Korean War and continues to this day.

As this narrative noted at the beginning, the 12,300 RAF aircrew members (4,300 at the AAF schools, including those at Maxwell and Gunter Fields; the 6,000 trained at the contractor BFTSs; and the 2,000 navigators trained at the Pan American Airways navigation school at Coral Gables, Florida) equaled about 10 percent of the over 131,000 number that the RAF trained through the BCATP during the war. Thus, the AAF-trained RAF aircrew members were almost inconsequential to the overall combat capability of the British and Dominion air forces during the war. In other words, the British and Dominion air forces would have had virtually the same level of combat capability without the AAF-trained pilots. Therefore, how did the AAF-trained RAF aircrew members really contribute to the war effort? First of all, the AAF flight training program served as a bridge between the startup of the BCATP in 1940 and the full production mode of the BCATP schools by mid-1942 (as noted, these schools produced 39,000 aircrew members in 1942 alone). With the BCATP in full production by mid-1942, the RAF no longer needed those undergoing training at the AAF schools, and as a result, notified the AAF that the RAF would be terminating the training of RAF flight cadets with Class 42-B, beginning on September 8, 1942 and graduating on February 26, 1943 (and incrementally curtailing the training at the contractor BFTSs).

In addition to serving as a bridge between the start-up of the flight training and the achievement of full training mode by the BCATP, the flight training by the AAF schools also demonstrated the desire of Roosevelt to do what he could legally do to help the beleaguered British, standing alone against the Nazis, given the prevalent isolationist and non-interventionist sentiment in the United States in early 1941 before America's entry in to the war. Roosevelt's position to help the British in their time of need also reflected the traditional affinity of many Americans toward their European cousins. In a relatively short time, the US government obtained the legal means to fund military aid (the Lend-Lease Act), including AAF flight training of RAF cadets in the United States, the establishment of six contracted flying schools for RAF flight trading (the BFTSs), and flight training of RAF flight cadets at the AAF flying schools of SEACTC/EFTC—all within six months. The quick response of the US government in general and the AAF in particular in establishing an extensive flight training plan for the RAF demonstrated the significant degree of American support for Britain in its time of need although the United States was still neutral and, in a more general sense, the US government's support to help potential and, after December 7, 1941, real allies.

The reasons for the establishment of a flight training program at the AAF schools for the Free French were quite different for the RAF. From June 1940, the Germans occupied about two-thirds of France until November 1942 when they occupied the rest of the country. Until mid-1943, the Free French government-in-exile in London, led by General de Gaulle, had relatively few operational military units, including a small air force. The Allied liberation of French North Africa in early 1943 provided the Free French with a base of operations and about 14,000 air force personnel, including about 1,400 pilots, eager to participate in the liberation of their country. However, at the time the Free French had few sources of flight training. As a result, the termination of the RAF flight training in February 1943 opened up training capacity in the AAF schools for the FFAF to begin flight training in June 1943. Although the AAF trained only 1,277 pilots by December 1945, these smaller numbers, compared to the number of RAF pilots who completed the AAF flight training program, had a greater overall effect on the wartime combat operations of the FFAF since it was a smaller air force than the RAF. As noted above, the FFAF had 1,500 pilots in mid-1943, and, since about 1,000 FFAF pilots were killed during war, a sizeable portion of the surviving pilots of the FFAF in late 1945 had to be AAF flight training graduates. Thus, these officers, in turn, had to form a sizeable portion of the cadre for the postwar French Air Force, the *Armee de l'Air*. As in the case of the RAF flight training, the AAF's training of FFAF aircrew demonstrated the ability and willingness of the US government to support its allies during wartime.

A second significant conclusion about the AAF training of both RAF and the FFAF flight cadets from June 1941 through November 1945 comes by comparing the American, RAF, and FFAF completion, elimination, and accident rates. The Americans, RAF, and FFAF flight cadets went through the same flight training phases, the same training curricula, the same non-duty experiences (for example, "square meals," honor code, and so on), and the same disciplinary procedures (more or less with some exceptions for the RAF and FFAF cadets) as the American flight cadets. During the flight training programs, senior American, British, and French military officials

5.1. In Oct 2009, members of the German Air Force Flying Training Center, Holloman AFB, New Mexico, commemorates completion of 55,555th flight hours by center aircraft by standing in a formation of five fives with 55 people in each five.

were rightly concerned about the "high" elimination and accident rates for their respective groups of cadets, especially the RAF senior officers in Washington and in Britain. As a result, some of these officials considered the training program a failure, and certainly many of the washed-out students were embittered by this negative experience. Still, in the end, in looking at the Americans, British, and French cadets as collective groups, they generally had statistically the same completion, elimination, and accident rates, despite the different backgrounds and cultural differences, and, thus, in that sense, the AAF training of RAF and Free French was an overall success.

One would expect that the French especially would have had a lower completion rate and higher elimination and accident rates than the Americans, given the language differences that produced problems in the actual training programs. These problems included the requirement to translate training manuals and training aids into French (which the EFTC still worked on in late 1944), problems in translating technical and idiomatic terms into the correct French, the lack of an adequate number of French-speaking American instructors, the need to use translators on the ground to go over preflight instructions and post-flight discussion of training flights, and the use of hand gestures between French students and American instructors during actual training flights. The more or less statistically same completion, elimination, and accident rates are testaments to the desire of the French flight cadets to succeed in their training program and their American instructors to ensure the French received the same quality flight training instruction as the Americans.

Finally, the AAF training of over 12,000 RAF cadets (4,360 from the AAF schools, over 6,000 from the BFTSs, and about 2,000 from the Pan American navigation school) and 4,113 FFAF aircrew members (plus another 4,600 from 27 other countries) represented a significant effort by the AAF to train aircrew for Allied air forces during World War II at the same that the AAF was training thousands of American flight cadets. The US military services quickly demobilized after September 1945, but the onset of the Cold War after the Berlin Airlift, the formation of the North Atlantic Treaty Organization (NATO) in April 1949, and the onset of the Korean War led the newly established Department of Defense (DoD) to institute new security assistance programs that would train hundreds of thousands of members of the armed forces, including the air forces, of allied nations across the globe. It is difficult, if impossible, to determine if the senior DoD leaders in the early 1950s looked back to the military training programs, such as the "Arnold Scheme," of World War II as models or prototypes for the various military training programs the US military services instituted during the Korean War. However, I suspect that some senior US military leaders did remember the success of these programs.[172]

Between 1950 and 2002, the last year that the Air Education and Training Command (AETC) tracked international students, 7,362 international students had graduated from the Air Force's undergraduate pilot training programs, usually as individuals or small groups.[173] In addition, the US Air Force provided land and airspace at Holloman AFB, New Mexico for the German Air Force (GAF) which constructed and has operated the GAF Tactical Training Center since May 1, 1996 (redesignated as the German Air Force Flying Training Center on July 1, 1999), reminiscent of the AAF support for the six BFTSs in the United States during World War II.[174]

5.2. Logo of the Euro-NATO Joint Pilot Training Program, Sheppard AFB, Texas, with the flags of the 13 nations (Belgium, Canada, Denmark, Germany, Greece, Italy, Netherlands, Norway, Portugal, Spain, Turkey, United Kingdom, and United States) who provide the staff, instructor pilots, and/or students.

Finally, since October 23, 1981, the Euro-NATO Joint Jet Pilot Training Program, a combined NATO training program at Sheppard AFB, Texas, in which US Air Force students participate, has trained 6,622 flight cadets from 13 participating NATO countries at Sheppard AFB, which has provided flight training to international military students since 1965.[175] From time to time, the Air Force also provided flight training to specific groups, such as the South Vietnamese and Laotian air forces during the Southeast Asian conflict.[176]

As a final comment, as noted earlier, it took several months for the leadership of the SEACT/EFTC to realize that the RAF and especially the FFAF cadets going through its flight training schools needed some level of formal exposure to American culture, history, and language (FFAF) before starting their actual flight training. As a result, SEACTC/EFTC instituted a preliminary course for the RAF and FFAF cadets (preflight school, RTC, French Reclassification Center) which, in addition to general military training, provided some education in American history, culture, and, for the French cadets, English. When the Department of Defense renewed providing flight training and professional military education for international students by 1950, the officials at the training centers and at Air University, Maxwell Air Force Base, eventually realized that these students also needed some level of preparation in American history, culture, and English before they began their training/education programs. According to Mr. Gregory

5.3. Col Lance Bunch, 80th Flying Training Wing commander, (l) and Italian Air Force Lt Col Michele Minonne, 469th Flying Training Squadron senior national representative, pose after completing Minonne's 5,000th hour of flight training in a T-38 Talon, Sheppard AFB, Texas, Jan 6, 2013, as part of the ENJPT program. Minonne has flown Italian fighter aircraft, such as the Fiat G91, AMX International, and the Aermacchi MB339 A/CDs, but has flown the majority of his hours as an instructor pilot in the T-37 "Tweety Bord" and the T-38. He has served as an instructor pilot since 1997.

L. Baker, Foreign Disclosure Officer, Air University, most, if not all, US military training and education installations require international students to achieve a specified score on the Test of English as a Foreign Language (TOEFL) as part of the selection process for a US military training or education program or attendance at the Defense Language Institute English Language Center, Lackland AFB, Texas, prior to their education or training program.[177] As the SEATC/EFTC learned from hindsight in training the FFAF cadets and Hines noted in several places in his monograph, the English language capability of international students was a very key (if not essential) factor in them performing well in their training programs, and the best method for teaching international students English was to have them complete formal English training before they started their training or education programs.[178]

Additionally, many US military training and education facilities have established special organizations to assist international military students and their families with arrival and in-processing, acclimatization to the installation and local community, and introductory programs related to their training or education programs to introduce international students to "America" before they begin their training or education program. For example, the International Office School (IOS) at Maxwell AFB and its predecessors have provided these services to international students selected to attend the Air Force's professional military education (PME) schools (Squadron Officer College [SOC], Air Command and Staff College [ACSC], and Air War College [AWC]) since 1945. Generally, international students report to Maxwell AFB two months before there formal education programs begin to go through preparatory courses that focuses on communication skills (speaking, giving briefings, and conversational English), American teaching methods (lecture, seminar discussion), and American history, government, and culture.[179] These preparatory programs acclimate the international students to their environment in central Alabama and significantly assist them in completing their education and training programs at Maxwell AFB.

In addition, the IOS, like many other military centers with international students, provides a formal sponsor program (Alabama Goodwill Ambassadors) through which volunteer families from the Montgomery area communities assist international students and their families (for students attending ACSC and AWC) to settle into the Montgomery community. They help the students to find housing, obtain driver's licenses, register cars as needed and enroll school-age children in local schools. In addition, the sponsor families host a number of social activities for the international families during their time at Air University so that the international students and their families receive a taste of American home life and culture. The sponsor program evolved over many years after 1953 when Air University leaders finally realized that the international students and their families needed such formal programs, versus the *ad hoc* activities such as those that the RAF and FFAF cadets experienced during their time in the United States, to help them acclimate to American society and culture so that they could perform at their peak level during their formal education programs.[180]

Appendices

Appendix 1. AAF Training of Foreign Nationals during World War II

Country	Number Trained	Country	Number Trained
Argentina	24	Mexico	447
Australia	55	Netherlands	532
Bolivia	46	New Zealand	1
Brazil	814	Nicaragua	3
Canada	110	Norway	2
Chile	50	Paraguay	2
China	2,238	Peru	52
Costa Rica	1	Poland	12
Cuba	36	Soviet Union	28
Czechoslovakia	1	Turkey	49
Ecuador	16	Union of South Africa	1
France	4,113	Uruguay	10
Great Britain	12,561	Venezuela	9
Haiti	7	Yugoslavia	68
Honduras	4		
Total			21,301

NOTE: This figure includes graduates of schools operated by the British and the Dutch with AAF assistance as well as those who graduated from AAF schools. Available statistics do not always agree, but this table (which includes both pilot and technical training) is approximately correct. (Note from source: These figures include foreign nationals who received training as pilots, navigators, flight engineers, bombardiers, and gunners.)

Source: Thomas Greer, "Other Training Programs," in Wesley F. Craven and James Lea Cate, eds., *The Army Air Forces in World War II, Vol. VI, Men and Planes* (Chicago: University of Chicago Press, 1955), p. 694.

Appendix 2. Lend-Lease Act, March 11, 1941

Further to promote the defense of the United States, and for other purposes.

Be it enacted by the Senate and House of Representatives of the United States of America in Congress assembled, That this Act may be cited as "An Act to Promote the Defense of the United States".

SEC. 2. As used in this Act –

(a) The term "defense article" means –

(1) Any weapon, munition. aircraft, vessel, or boat;

(2) Any machinery, facility, tool, material, or supply necessary for the manufacture, production, processing, repair, servicing, or operation of any article described in this subsection;

(3) Any component material or part of or equipment for any article described in this subsection;

(4) Any agricultural, industrial or other commodity or article for defense.

Such term "defense article" includes any article described in this subsection: Manufactured or procured pursuant to section 3, or to which the United States or any foreign government has or hereafter acquires title, possession, or control.

(b) The term "defense information" means any plan, specification, design, prototype, or information pertaining to any defense article.

SEC. 3. (a) Notwithstanding the provisions of any other law, the President may, from time to time, when he deems it in the interest of national defense, authorize the Secretary Of War, the Secretary of the Navy, or the head of any other department or agency of the Government –

(1) To manufacture in arsenals, factories, and shipyards under their jurisdiction, or otherwise procure, to the extent to which funds are made available therefor, or contracts are authorized from time to time by the Congress, or both, any defense article for the government of any country whose defense the President deems vital to the defense of the United States.

(2) To sell, transfer title to, exchange, lease, lend, or otherwise dispose of, to any such government any defense article, but no defense article not manufactured or procured under paragraph (1) shall in any way be disposed of under this paragraph, except after consultation with the Chief of Staff of

the Army or the Chief of Naval Operations of the Navy, or both. The value of defense articles disposed of in any way under authority of this paragraph, and procured from funds heretofore appropriated, shall not exceed $1,300,000,000. The value of such defense articles shall be determined by the head of the department or agency concerned or such other department, agency or officer as shall be designated in the manner provided in the rules and regulations issued hereunder. Defense articles procured from funds hereafter appropriated to any department or agency of the Government, other than from funds authorized to he appropriated under this Act, shall not be disposed of in any way under authority of this paragraph except to the extent hereafter authorized by the Congress in the Acts appropriating such funds or otherwise.

(3) To test, inspect, prove, repair, outfit, recondition, or otherwise to place in good working order, to the extent to which funds are made available therefor, or contracts are authorized from time to time by the Congress, or both, any defense article for any such government, or to procure any or all such services by private contract.

(4) To communicate to any such government any defense information pertaining to any defense article furnished to such government under paragraph (2) of this subsection.

(5) To release for export any defense article disposed of in any way under this subsection to any such government.

(b) The terms and conditions upon which any such foreign government receives any aid authorized under subsection (a) shall be those which the President deems satisfactory, and the benefit to the United States may he payment or repayment in kind or property, or any other direct or indirect benefit which the President deems satisfactory.

(c) After June 30, 1943, or after the passage of a concurrent resolution by the two Houses before June 30, 1943, which declares that the powers conferred by or pursuant to subsection (a) are no longer necessary to promote the defense of the United States, neither the President nor the head of any department or agency shall exercise any of the powers conferred by or pursuant to subsection (a) except that until July 1, 1946, any of such powers may be exercised to the extent necessary to carry out a contract or agreement with such a foreign government made before July 1,1943, or before the passage of such concurrent resolution, whichever is the earlier.

(d) Nothing in this Act shall be construed to authorize or to permit the authorization of convoying vessels by naval vessels of the United States.

(e) Nothing in this Act shall be construed to authorize or to permit the authorization of the entry of any American vessel into a combat area in violation of section 3 of the Neutrality Act of 1939.

SEC. 4 All contracts or agreements made for the disposition of any defense article or defense information pursuant to section 3 shall contain a clause by which the foreign government undertakes that it will not, without the consent of the President, transfer title to or possession of such defense article or defense information by gift, sale, or otherwise, or permit its use by anyone not an officer, employee, or agent of such foreign government.

SEC. 5. (a) The Secretary of War, the Secretary of the Navy, or the head of any other department or agency of the Government involved shall when any such defense article or defense information is exported, immediately inform the department or agency designated by the President to administer section 6 of the Act of July 2, 1940 (54 Stat. 714). of the quantities, character, value, terms of disposition and destination of the article and information so exported.

(b) The President from time to time, but not less frequently than once every ninety days, shall transmit to the Congress a report of operations under this Act except such information as he deems incompatible with the public interest to disclose. Reports provided for under this subsection shall be transmitted to the Secretary of the Senate or the Clerk of the House of representatives, as the case may be, if the Senate or the House of Representatives, as the case may be, is not in session.

SEC. 6. (a) There is hereby authorized to be appropriated from time to time, out of any money in the Treasury not otherwise appropriated, such amounts as may be necessary to carry out the provisions and accomplish the purposes of this Act.

(b) All money and all property which is converted into money received under section 3 from any government shall, with the approval of the Director of the Budget, revert to the respective appropriation or appropriations out of which funds were expended with respect to the defense article or defense information for which such consideration is received, and shall be available for expenditure for the purpose for which such expended funds were appropriated by law, during the fiscal year in which such funds are received and the ensuing fiscal year; but in no event shall any funds so received be available for expenditure after June 30, 1946.

SEC. 7. The Secretary of War, the Secretary of the Navy, and the head of the department or agency shall in all contracts or agreements for the disposition of any defense article or defense information fully protect the rights of all citizens of the United States who have patent rights in and to any such article or information which is hereby authorized to he disposed of and the payments collected for royalties on such patents shall be paid to the owners and holders of such patents.

SEC. 8. The Secretaries of War and of the Navy are hereby authorized to purchase or otherwise acquire arms, ammunition, and implements of war produced within the jurisdiction of any country to which section 3 is applicable, whenever the President deems such purchase or acquisition to be necessary in the interests of the defense of the United States.

SEC. 9. The President may, from time to time, promulgate such rules and regulations as may be necessary and proper to carry out any of the provisions of this Act; and he may exercise any power or authority conferred on him by this Act through such department, agency, or officer as be shall direct.

SEC. 10. Nothing in this Act shall be construed to change existing law relating to the use of the land and naval forces of the United States, except insofar as such use relates to the manufacture, procurement, and repair of defense articles, the communication of information and other noncombatant purposes enumerated in this Act.

SEC 11. If any provision of this Act or the application of such provision to any circumstance shall be held invalid, the validity of the remainder of the Act and the applicability of such provision to other circumstances shall not be affected thereby.

Approved, March 11, 1941.

Source: "Lend Lease Act, 11 March 1941," Naval Historical Center, Washington, DC, at http://www.history. navy.mil/faqs/faq59-23.htm, accessed May 13, 2013. Originally from *Public Laws*, Part 1 of *United States Statutes at Large Containing the Laws and Concurrent Resolutions Enacted During the First Session of the Seventy-Seventh Congress of the United States of America, 1941-1942, and Treaties, International Agreements Other than Treaties, and Proclamations*, Vol. 55 (Washington: Government Printing Office, 1942), pp. 31-33.

Appendix 3.
Southeast Training Center Flying Schools, May 1, 1943

HEADQUARTERS: MAXWELL FIELD, AL

6 PRIMARY FLYING SCHOOLS

Van de Graaf Field, Tuscaloosa, AL*
Carlstrom Field, Arcadia, FL
Contract School, Lakeland, FL*
Contract School, Albany, GA*
Souther Field, Americus, GA*
Moody Field, Valdosta, GA

11 BASIC FLYING SCHOOLS

Courtland AAF, Courtland, AL
Newport AAF, Newport, AL
Gunter Field, Montgomery, AL
Greenwood AAF, Greenwood, MS
Walnut Ridge AAF, Walnut Ridge, AL
Malden AAF, Malden, MO
Bainbridge AAF, Bainbridge, AL
Shaw Field, Sumter, SC
Cochran Field, Macon, GA
Tuskegee AAF, Tuskegee, AL (Basic and Advanced)
Greenville AAF, Greenville, MS

4 ADVANCED SINGLE ENGINE SCHOOLS

Craig Field, Selma, AL
Napier Field, AL
Marianna AAF, Marianna, FL
Spence Field, GA

7 Advanced Two-Engine Schools

Blytheville AAF, Blytheville, AR
Stuggart AAF, Stuggart, AL
Freeman AAF, Seymour, IN
Moody Field, Valdosta, GA
Turner Field, Albany, GA
George Field, Lawrenceville, IL
Columbus AAF, Columbus, MS

3 Combat Crew (4-Engine) and Bombardier Schools

Hendricks Field, Sebring,FL
Lockbourne AAB, Columbus, OH
Smyrna AAF, Smyrna, TN

5 Gunnery Schools (Flexible and Fixed)

Buckingham AAF, Fort Myers, FL
Naples AAF, Naples, FL
Tyndall Field, Panama City, FL
Apalachicola AAF, Apalachicola, FL
Eglin Field, Valparaiso, FL

2 Special Schools

Maxwell Field, Montgomery, AL
Jackson AAB, Jackson, MS

Source: Appendix II, Lloyd H. Cornett Jr., *Air Training Command A 30-Year View 1943-1974* (Randolph AFB, TX: HQ Air Training Command, History and Research Division, 1974), pp. 43-44.

Appendix 4. Training Schools for RAF Flight Crew in the USA, 1942-44

Type of School	'42	'43	'44
British Flying Training Schools[1]	6	6	5
US Army Air Force Schools (Arnold Plan)[2]			
Primary	5		
Basic Training	2		
Advanced Training	3		
Advanced Flying		4	
US Naval Flight Schools (Towers Plan)[3]			
Pilots	1	1	1
Radio Operators/ Air Gunners. Pilots and Radio Operators. Air Gunners Advanced	1		
Pilots Advanced		1	1
Refresher Schools[4]	3		
Pan-American Airways School[5]			
Navigators	1		

1. Civilian-operated schools under British contract. Provided 9-weeks' basic and 18-weeks' advanced flight training. Curriculum equivalent to RAF flight training school.
2. AAF flight schools for RAF cadets, using AAF curricula, for 9 weeks at each type of school. The AAF closed the primary and basic flying schools in October and December 1942 and the advanced schools in February 1943.
3. US Navy schools for RAF pilots, radio operators, and Fleet Air Arm pilots. Complete course was about six months long.
4. Provided refresher training for Americans recruited to serve in RAF (Eagle Squadrons)
5. Provided navigation training for RAF personnel

Source: Appendix C9, Training Schools for RAF Flight Crew in the USA, 1941-4, John Golley, *Aircrew Unlimited The Commonwealth Air Training Plan during World War 2* (Sparkford, UK: Patrick Stephens Limited, p. 178.

Appendix 5. Phases of AAF Flight Training, c. 1943-44

The official histories of the SEACTC/EFTC, Maxwell Field, and Gunter Fields provided details for the changes in length and curricula for each type of school from 1941 to 1945. This chart depicts the status in May 1944 of the standard AAF flight training program, which illustrated, in general, the overall AAF flight training program, in the first two years of the war.

Replacement Training Center: established in February 1941; redesignated as Pre-Flight School on 30 April 1942; originally four weeks; expanded to eight weeks in Mar 1943; curriculum varied from school to school until AAF standardized it in April 1943, final revision of May 1944 extended pre-flight school to ten weeks.

Curriculum: military discipline, dress and courtesies, close order drill, physical fitness, mathematics, target identification, photography, meteorology, map reading, aircraft and naval vessel identification, radio code, chemical warfare, small arms familiarization.

Consisted of two schools, the Air Corps Replacement Center (Pilot) for flying cadets and the Air Corps Replacement Center (Air Crew) for flying cadets who failed pilot training or selected as bombardiers and navigators.

Locations for RAF cadets: Maxwell Field, AL, Sep 1941 – Feb 1942; Turner Field, GA, Mar 1942-Feb 1943

Location for French cadets: Craig Field, AL. NOTE: Not a "true" RTC but served the essential function to in-process arriving French cadets

Phase 1: Elementary/Primary: Weeks 1–9

Curriculum: Phase 1: general operation of light aircraft, proficiency in landing techniques, recovery from stalls and spins; Phase 2: precision control through flying standard courses or patterns; Phase 3: high proficiency in landing approaches and landings; Phase 4: acrobatics, loops, Immelmann turns, slow rolls, snap rolls. Completion was soloing after at least 175 landings and demonstration of flying abilities without an instructor.

Locations for RAF Cadets: Arcadia, FL; Tuscaloosa, AL; Lakeland, FL; Albany, GA; Americus, GA; Camden, SC

Location for French cadets: Tuscaloosa, AL; Orangeburg, SC,

PHASE 2: BASIC: WEEKS 10-18

Curriculum: ability to fly a more advanced trainer (greater weight, power, and complexity); instrument, night, formation, and cross-county flying; more precision and smoothness in flying, repetition of maneuvers; included Link trainer instruction.

During basic school, a flying board determined which AAF cadets would attend advanced single-engine schools and which would attend advanced 2-engine schools.

Locations for RAF cadets: Gunter, AL; Macon, GA;

Location for French cadets: Gunter, AL

PHASE 3: ADVANCED SINGLE-ENGINE (PURSUIT): WEEKS 19-31

Curriculum: transition, instrument, navigation, formation and acrobatics; Link trainer instruction; fixed aerial gunnery; increasing emphasis on formation flying

Locations for RAF cadets: Craig Field, AL; Maxwell, AL (See narrative for narrative about the status of the advanced school at Maxwell.)

Location for French cadets: Craig Field, AL

Advanced 2-Engine (Bombers, Transport, 2-engine fighters): Weeks 19-42

Curriculum: transition, instrument, navigation, formation and acrobatics; Link trainer instruction; flexible aerial gunnery; increasing emphasis on formation flying

An AAF Aviation Cadet received his pilot's wings and commission as a second lieutenant at graduation from his respective advanced school.

Locations for RAF cadets: Turner Field, GA

Location for French cadets: Turner Field, GA

Once the RAF and Free French cadets graduated from their respective advanced schools, they returned to Britain or North Africa, respectively, for transition and operational training in a specific aircraft, depending on the needs of the RAF and FFAF. The RAF did not go through transition and operational training under the Arnold Plan. Some of the FFAF pilots

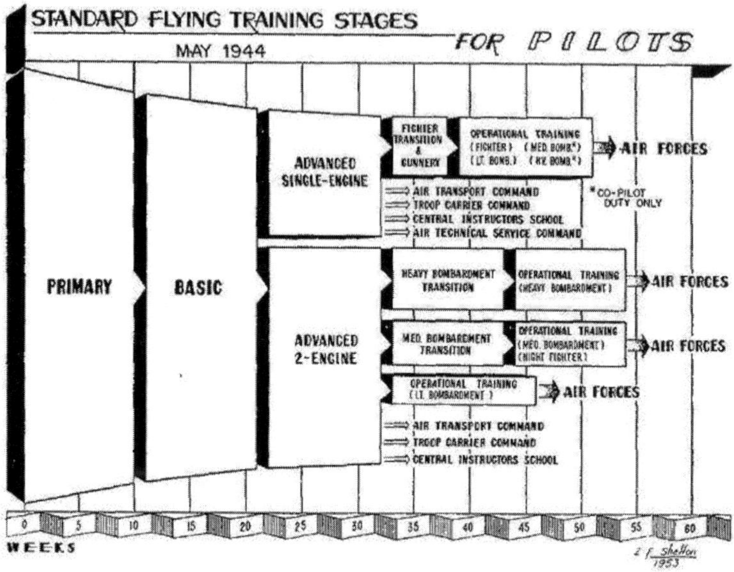

6.1. Diagram of US Army Air Forces flight training program, mid-1943.

after graduating from their advanced schools did receive some P-40 and B-25 transition training prior to leaving for their operational assignments in North Africa. Since this training did not occur at Maxwell or Gunter Fields, I did not present the detailed curricula for these two phases of flight training here. See Thomas H. Greer, "Combat Crew and Unit Training," Chapter 8, Craven and Cate, pp. 600-28, for specifics about AAF transition and operational training.

Phase 4: Transition: training in specific types of combat aircraft

Pursuit: Weeks 32-38

Heavy and Medium Bombardment: Weeks 32-43

Operational (Light Bombardment): Weeks 32-46

Phase 5: Operational Training

Training in specific type of aircraft to which pilot was assigned in an operational area before first combat mission. Length depended on type of aircraft and unit of assignment.

Source: "New Center Opens Today," *Training News*, Sep 6, 1941, pp. 1, 8. "Cadets Shifted at Many Schools in S.E.A.C.T.C.," *Training News*, Nov 8, 1941, p. 1, 2. Thomas H. Greer, "Individual Training of Flying Personnel" (Chapter 7), in Wesley F. Craven and James Lea Cate, eds., *The Army Air Forces in World War II, Vol. VI, Men and Planes* (Chicago: The University of Chicago, 1955), pp. 557-75. The chart is on p. 568.

Appendix 6a. RAF Pilot Trainees at AAF SEACTC Schools 42-A–42-G

Class	42-A	42-B	42-C	42-D	42-E	42-F	42-G
Entered Primary	550	549	620	640	745	747	739
Start Date Primary	06/07/41	07/26/41	08/27/41	10/04/41	11/06/41	12/17/41	01/11/42
Held Over Next Class	10	12	13	13	16	14	21
Eliminated	237	219	214	239	266	312	242
Killed	1	1	2	2	0	1	0
Grad. Pri./% Grad.	302/56	327/60	403/65	399/62	476/64	436/58	490/67
Start Date Basic	08/23/41	10/02/41	11/06/41	12/18/41	12/01/41	02/1942	03/30/42
Held Over Next Class	4	4	9	11	13	24	
Eliminated	11	11	26	47	60	43	21
Killed	0	1	0	2	2	8	72
Grad. Basic/% Grad.	287/96	315/96	372/93	348/88	412/87	373/88	9
Start Date Advanced	11/05/41	12/15/41	01/11/41	02/20/41	03/17/42	05/01/42	06/03/42
Held Over Next Class	7	8	13	4	13	6	13
Eliminated	2		7	7	14	28	14
Killed	8	5	4	1	1	2	9
Grad. Adv./ & Grad.	270/96	300/96	356/96	342/96	674/93	364/96	376/93
Grad. Date Adv.	01/03/42	02/07/42	03/06/42	04/24/42	05/17/42	07/03/42	08/05/42
Percent Completed	49.1	54.6	57.2	53.4	50.2	48.7	50.9

Appendix 6b. RAF Pilot Trainees at AAF SEACTC Schools 42-H–43-C

RAF Pilot Trainees at AAF SEACTC Schools 42-H – 43-C								
Class	42-H	42-I	42-J	42-K	43-A	43-B	43-C	Total
Entered Primary	754	501	493	505	515	502	11	7860
Start Date Primary	02/22/42	03/30/42	06/01/42	06/03/42	07/05/42	09/08/42		
Held Over Next Class	27	18	20	16	17	11	0	208
Eliminated	252	172	172	135	133	95	11	2699
Killed	0	0	0	1	1	0	0	9
Grad. Pri./% Grad.	496/66	338/67	319/65	373/73	380/74	410/81		5140/65
Start Date Basic	05/01/42	06/30/42	07/05/42	08/06/42	09/08/41	10/13/42	11/17/42	
Held Over Next Class	17	10	10	21	9	8	8	161
Eliminated	71	37	21	40	37	44	2	522
Killed	4	2	2	3	0	0	0	33
Grad. Basic/% Grad.	427/85	306/89	296/92	319/88	355/91	357/87	6/75	4,578/89
Start Date Advanced	07/05/42	08/07/42	09/08/42	10/11/42	11/12/42	12/13/42	01/21/43	
Held Over Next Class	10	9	8	9	9	7	0	116
Eliminated	18	14	11	14	10	16	3	173
Killed	1	2	1	0	1	1	0	36
Grad. Adv./ & Grad.	415/96	294/95	281/95	308/97	338/95	342/95	10/83	4,360/95
Grad. Date Adv.	09/06/42	10/09/42	11/10/42	12/13/42	01/14/43	02/26/43	03/25/43	
Percent Completed	55.9	58.7	57.0	57.0	65.6	68.0	38.5	55.5

Appendix 7a. Free French Pilot Trainees at AAF EFTC Schools 44-A–44-H

Free French Pilot Trainees at AAF EFTC Schools 44-A – 44-H								
Class	44-A	44-B	44-C	44-D	44-E	44-F	44-G	44-H
Entered Primary	104	105	97	105	129	133	138	0
Start Date Primary	07/02/43	07/29/43	09/01/43	10/03/43	11/04/43	12/06/43	01/16/44	02/14/44
Held Over Next Class	3	4	7	5	6	6	1	0
liminated	26	35	18	34	35	43	34	0
Killed	0	0	0	0	0	0	0	0
Grad. Pri./% Grad.	75/74.2	69/66.3	76/80.9	73/68.2	93/72.7	90/67.6	104/75.4	0
Start Date Basic	09/01/43	10/03/43	11/05/43	12/07/43	01/17/44	Note 3	03/27/44	04/22/44
Held Over Next Class	2	0	7	5	8	6	11	8
Eliminated	1	5	5	5	14	12	9	30
Killed	0	1	0	0	0	0	0	
Grad. Basic/% Grad.	72/98.6	65/91.5	64/92.8	70/93.3	77/84.6	80/87.0	89/90.6	1
Start Date Advanced	11/03/43	12/07/43	01/16/44	02/10/44	03/03/44	Note 3	Note 3	06/30/44
Held Over Next Class	3	7	3	2	8	11	7	4
Eliminated	4	46	6	5	11	11	17	3
Killed	0	2	0	1	0	0	0	0
Grad. Adv./ & Grad.	64/74.2	63/56.8	61/91.0	67/91.7	65/85.5	74/92.5	81/82.6	6
Grad. Date Adv.	Note 3	Note 3	Note 3	04/15/44	05/23/44	06/27/44	08/04/44	09/08/44
Percent Completed	61.5	60.0	62.9	63.8	50.4	55.6	58.7	

Appendix 7b. Free French Pilot Trainees at AAF EFTC Schools 44-I–45-E

FREE FRENCH PILOT TRAINEES AT AAF EFTC SCHOOLS 44-I– 45-E									
Class	44-I	44-I–1	44-J	44-K	45-A	45-B	45-C	45-D	45-E
Entered Primary	132	132	133	150	150	150	152	150	152
Start Date Primary	03/27/44	03/13/44	04/23/44	05/26/44	06/28/44	08/07/44	09/11/44	11/20/44	12/31/44
Held Over Next Class	2	6	3	10	17	5	19	19	18
Eliminated	35	18	37	47	55	56	55	43	60
Killed	0	0	0	1	0	0	1	0	1
Grad. Pri./% Grad.	101/74.3	87/68.5	94/71.2	95/66.4	88/61.5	106/65.4	82/59.4	107/71.3	90/59.6
Start Date Basic	05/26/44	05/26/44	06/30/44	08/07/44	09/11/44	11/23/44	12/31/44	02/07/45	03/18/45
Held Over Next Class			8	2	4	10	5	9	7
Eliminated	23	29	21	25	31	33	17	21	17
Killed	0	0	1	0	0	3	0	0	0
Grad. Basic/% Grad.	73/76.0	64/68.8	72/76.5	74/73.2	56/64.3	66/66.0	70/80.5	82/79.6	75/81.5
Start Date Advanced	08/07/44	08/05/44	09/09/44	11/23/44	12/31/44	02/07/44	03/18/44	03/18/45	05/26/45
Held Over Next Class	5	8	6	2	56	6	4	9	7
Eliminated	15	9	7	17	8	8	10	21	17
Killed	0	2	3	2	0	0	0	0	0
Grad. Adv./ & Grad.	61/80.2	60/90.9	62/86.1	59/75.6	45/84.9	55/87.3	62/86.1	82/79.6	75/81.5
Grad. Date Adv.	11/20/44	Note 3	12/23/44	02/01/45	03/11/45	Note 3	05/23/45	06/27/45	08/04/45
Percent Completed	46.2	45.5	46.6	39.3	30.0	36.7	40.8	54.7	50.0

Appendix 7c. Free French Pilot Trainees at AAF EFTC Schools 45-F–H

Free French Pilot Trainees at AAF EFTC Schools 45-F – 45-H				
Class	45-F	45-G	45-H	Total
Entered Primary	152	152	150	2,564
Start Date Primary	02/07/45	03/18/45	04/18/45	
Held Over Next Class	14	16	8	169
Eliminated	56	47	40	774
Killed	0	0	0	3
Grad. Pri./% Grad.	100/64.1	103/68.7	118/74.7	1,568/61
Start Date Basic	06/30/45	05/26/45	06/30/45	
Held Over Next Class	6	6	7	121
Eliminated	27	10	17	352
Killed	1	1	0	6
Grad. Basic/% Grad.	73/72.3	93/90.3	100/85.5	1,415/90
Start Date Advanced	06/30/45	08/07/45	9/11/45	
Held Over Next Class	4	6	5	163
Eliminated	10	9	17	251
Killed	0	1	0	11
Grad. Adv./ & Grad.	65/86.7	93/90.3	83/83.0	1,277/90
Grad. Date Adv.	09/08/45	10/16/45	11/20/45	
Percent Completed	42.8	61.2	55.3	49.8

A detailed review of the EFTC histories Jun 1943 to Dec 1945 will provide slight variations in absolute numbers and percentages than noted here. The number graduated also includes students held over from a previous class who eventually graduated.

Note 1: All of the French students went to the Gunter Basic School. Its graduates went to the advanced single-engine school, Craig Field, AL, or the advanced two-engine school at Turner Field, GA.

Note 2: There were no French students in the primary school for Class 44-H. The basic school for Class 44-H had 12 students held over from the previous class, 8 of which were held over for the next basic class, giving this class a net class 4 of which 3 were eliminated during the class, leaving one to graduates. The advanced school for Class 44-H had five students to enter the class who joined eight students held over from the previous class. Four were held over, giving the advanced class a net of nine students. Three were eliminated, leaving six to graduate.

Note 3. The source below did not provide these dates.

Note 4. The percentages in the last column represent the percentage of FFAF cadets who completed the entire AAF flight training program; calculated by author.

Source: AAFFTC Pilot Class Histories, Total Training, Classes 44-A through 45-H, *EFTC History, 1 Sep 1945 – 15 Dec 1945*, Vol. 5, Appendix, AFHRA Call # 222.01,V. 5, Doc # 00146165.

Appendix 8. AAF Training Aircraft Flown by RAF/FFAF Cadets

Primary School: Stearman PT-13

PT-13 assigned to a SEACTC primary school

Source of Photograph: National Museum of the United States Air Force. Specifications: National Museum of the United States Air Force (NMUSAF), "Stearman PT-13" http://www.nationalmuseum.af.mil/Upcoming/ArticleDisplay/tabid/466/Article/579880/13-cool-facts-about-the-world-war-ii-pt-13.aspx.

Basic School through late 1943: Vultee BT-13 Valiant

Gunter BT-13 assigned to Gunter Field Basic School

Source of Photograph: AU/HO Photograph Files.

Basic School after Mid-1943 and Advanced Single-Engine School:

North American AT-6

AT-6 Texans, Maxwell Advanced

Single-Engine School, c. 1942

Source of Photograph: AU/HO Photograph Files. Specifications: Hill Air Force Base Museum, "AT-6A Texan," http://www.hill.af.mil/library/ factsheets/factsheet.asp?id=5665, accessed Jun 9, 2013.

Advanced Two-Engine School:

Cessna AT-17 Bobcat

Source of Photograph: NMUSAF, "Cessna C-78B Bobcat," at http://www.nationalmuseum.af.mil/factsheets/factsheet.asp?id=516, accessed 15 Jun 2013. Specifications: Warbird Alley, "Cessna T-50/UC-78 Bobcat," at http://www. warbirdalley.com/uc78.htm

Curtiss AT-9 Jeep/Fledgling
Curtiss AT-9 assigned to Maxwell Field Advanced Two-Engine School, c. 1943

Source of Photograph: AU/HO Photograph Files.

AT-10 Wichita

Specifications: NMUSAF, "Beech A-10 Wichita," http://www.nationalmuseum.af.mil/factsheets/factsheet.asp?id=483, accessed Jun 13, 2013.

North American AT-24 Mitchell
(Trainer version of B-25 Mitchell)

Source of Photograph: National Museum of the United States Air Force. Specifications: National Museum of the United States Air Force (NMUSAF), "North American TB-25" http://www.nationalmuseum.af.mil/Upcoming/Photos.aspx?igphoto=2000549064.

Appendix 9. RAF 1st Lt. John Clark

6.2. RAF Flight Cadet John Clark attended Class 42-E at the basic school at Gunter Field and then the advanced school at Maxwell. After graduation from the advanced school at Maxwell Field, he became a flight instructor at the Gunter basic school.

In Jan 1941, Clark joined the RAF and, after preliminary training in Britain, arrived at Albany, GA, for Class 42-E on Nov 6, 1941, for primary training. After graduation from primary training, Clark arrived at Gunter Field on Jan 9, 1942, for the basic school and Maxwell for the advanced school. After graduating and receiving his officer's commission in May 1942, the RAF Delegation assigned him to Gunter Field as an instructor pilot at the basic school. In March 1942, he met Jean Harmon at the Morrison Cafeteria in Montgomery and married her in Sep. In spring 1943, Clark returned to Britain as a flight instructor at various RAF bases. Soon afterward, his wife joined him as an employee of the American Red Cross. The Clarks returned to the United States separately by summer 1945, and Mrs. Clark gave birth to their only child Rose in Sep 1945. They settled in Birmingham where John became a banker and eventually moved to Selma, where he retired in 1992. John died on Oct 18, 2013, and his daughter, whom I met in early 2014, donated his World War II documents to the Air Force Historical Research Agency, Maxwell AFB, and several uniform items and other memorabilia to Air University (AU). Using these artifacts, AU created an exhibit to the RAF flight cadets who trained at Maxwell and Gunter Fields during World War II.[181]

6.3. RAF Flight Cadet John Clark in a BT-13 during his basic flight training at Gunter Field, summer 1942.

6.4. Author and Ms. Rose Koch, daughter of RAF First Lieutenant John Clark, next to exhibit, dedicated to World War I IRAF flight training program, at Air University, Maxwell AFB, Alabama, on November 9, 2015. The exhibit displays uniforms and memorabilia of John Clark.

United States Army

Southeast Air Corps Training Center

Be it known that LAC John Clark,
Royal Air Force, has satisfactorily completed the course of instruction prescribed for Pilots in the Southeast Air Corps Training Center. In testimony whereof and by virtue of vested authority I do confer upon him this

DIPLOMA

Given at Maxwell Field this 16th day of May in the year of our Lord one thousand nine hundred and forty-two.

Attest:

C. W. Davies
Commandant, Advanced Flying School

Major General, U. S. Army,
Commanding, Southeast Air Corps Training Center

6.5. Graduation diploma for RAF Flight Cadet John Clark from the Maxwell advanced school.

Bibliography

Primary Sources

Document Collections

Air Force Historical Research Agency (AFHRA), Maxwell AFB, AL. Folders containing original documents and copies of original documents that pertain to the origins, implementation, and termination of the AAF flight training programs for RAF and FFAF flight cadets in the United States, including Maxwell and Gunter Fields, between March 1941 and November 1945. Each folder has a call number and a document number (for example, AFHRA Folder Call # 220.7271-4 British 1941, Doc # 00145958). The individual documents in each folder were normally (but not necessarily) arranged chronologically. See footnotes for specific folders used.

Air University Directorate of History (AU/HO), Maxwell AFB, AL. Folder containing original documents and copies of original documents that pertain to the RAF and French cadet graves in Oakwood Cemetery, Montgomery, Alabama

Franklin D, Roosevelt Presidential Library, Hyde Park, NY. Box 360, Papers of Harry L. Hopkins, containing correspondence related to the initial use of the Lend-Lease Act to fund RAF flight training in the United States and the origins of the RAF flight training program, March-June 1941. Referred to as Hopkins Papers.

"Roosevelt Speeches," *The American Presidency Project*, at http://www.presidency.ucsb.edu/ franklin_roosevelt.php, accessed May 29, 2013

"Selected Speeches of Winston Churchill," Churchill Centre and Museum at the Churchill War Rooms, at http://www.winstonchurchill.org/learn/speeches/speeches-of-winston-churchill/97-give-us-the-tools, accessed May 13, 2013

Memoirs

Arnold, Henry. H. *Global Mission*. New York: Harper & Brothers, 1949.

Other

Huston, John W. Maj Gen, USAF Retired, editor. *American Airpower Comes of Age General Henry H. "Hap" Arnold's World War II Diaries*. Maxwell AFB, AL: Air University Press, 2002. This two-volume work is a combination of selected excerpts from Arnold's diaries with historical commentary and narrative by

Gen Huston.

Official Histories

The official histories, used in this work, are located in the AFHRA, Maxwell AFB, AL. Specific volumes used are denoted by title, AFHRA Call # and Document #; for example, *Gunter Field History* (short title, see entries below), *1 Jul 1945 – 31 Aug1945*, pp. 7, 9-11, AFHRA Call # 283.81-10 Jul-Aug 45, Doc #00172974.

AAF Flying Training Command: *History of the AAF Flying Training Command and Its Predecessors 1 January 1939 to 7 July 1943*. Fort Worth, TX: HQ AAF Training Command, Mar 1, 1945. Referred to as "AAF FTC History" after first usage.

Gunter Field: *Station History Army Air Forces Pilot School (Basic) Gunter Field Alabama*, noted volumes with a beginning date of 7 December 1941 through December 31, 1945. Volumes, dated Dec 7, 1941 through Jun 30, 1944, are in the Air University Directorate of History (AU/HO), Maxwell AFB, AL; volumes, dated Jul 1, 1944 through Dec 30, 1945 are in the AFHRA. Referred to as "Gunter Field History" after first usage.

Maxwell Field: *History of Maxwell Field*; *Station History Maxwell Field, Alabama*; and *History of 2132nd AAF Base Unit Pilot School, Specialized, VHB [Very Heavy Bomber) and Maxwell Field, Alabama*; beginning date of 7 December 1941 through December 31, 1945. Referred to as "Maxwell Field History" after first usage.

Southeast Air Corps Training Center/Army Air Forces Eastern Flying Training Command: *History of Army Air Forces Eastern Flying Training Command*, noted volumes with a beginning date of January 1, 1939 through December 15, 1945. Referred to as "AAF EFTC History" after first usage.

Government Documents

Executive Order 7964 (now 20 USC 221), "Instruction of Citizens of the American Republics at Educational Institutions and Schools Maintained and Administered by the Government of the United States," Aug 29, 1938, and later modified, at http://law.onecle.com/uscode/ 20/221.html, accessed Jun 9, 2014.

"Lend Lease Act, 11 March 1941," Naval Historical Center, Washington, DC, at http://www. history.navy.mil/faqs/faq59-23.htm, accessed May 13, 2013.

Stettinius Jr., Edward R. *Report to the 78th Congress on Lend-Lease Operations from the Passage of the Act, March 11, 1941, to December 31, 1942*, Jan 25, 1943.

Secondary Sources

Books

Air University Directorate of History (AU/HO). *Cradle of Airpower Education A Short History of The Air University, Maxwell Air Force, and the 42nd Air Base Wing*. Maxwell AFB, AL: Air University Directorate of History, June 2012.

Buell, Thomas B.. Bradley, John N. and Bradley, John H. *The Second World War: Europe and the Mediterranean*. Garden City, NY: Square One Publishers, 2002.

Butler, Phil and Hagedorn, Dan. *Air Arsenal North America Aircraft for the Allies 1938-1945 Purchases and Lend-Lease*. Hinckley, UK: Midland Publishing, 2004.

Christienne, Charles and Lossarague, Pierre. *A History of French Military Aviation*. Washington, DC: Smithsonian Institution Press, 1986.

Clifton, William M., 80th Flying Training Wing Historian, "Background Paper on International Undergraduate Pilot Training at Sheppard Air Force Base, 1966-2013," Jun 11, 2013. Provided to author by Mr. Clifton.

Collins, James Lawton, Brig Gen, US Army. *Vietnam Studies: The Development and Training of The South Vietnamese Army, 1950-1972*. Washington, DC: US Army, 1991.

Craven, Wesley F. and Cate, James Lea, editors. *The Army Air Forces in World War II, Vol. VI, Men and Planes*. Chicago: The University of Chicago, 1955.

Dunmore, Spencer. *Wings For Victory*. Toronto: McClelland and Stewart, 1994.

Ennels, Jerome A. and Newton, Wesley Phillips. *The Wisdom of Eagles A History of Maxwell Air Force Base*. Montgomery, AL: Black Belt Press, 1997.

Gerhart, Eugene C. *America's Advocate: Robert H. Jackson*. Indianapolis: The Bobbs-Merrill, Inc., 1958.

Golley, John. *Aircrew Unlimited The Commonwealth Air Training Plan during World War2*. Sparkford, UK: Patrick Stephens Limited, 1993.

Guinn, Gilbert S. *The Arnold Scheme British Pilots The American South and Allies' Daring Plan*. Charleston, SC: The History Press, 2008.

Hines, Jay E. *History of Foreign Training in ATC [Air Training Command] 1941-1976*. Randolph AFB, TX: History and Research Division, Office of the Chief of Staff, Headquarters Air Training Command, [1977].

Manning, Thomas A. *History of Air Education and Training Command, 1942–2002*. Randolph AFB, TX: Headquarters Air Education and Training Command Office of History and Research, Headquarters, 2005.

Roll, David L. *The Hopkins Touch Harry Hopkins and the Forging of the Alliance to Defeat Hitler*. Oxford, UK: Oxford University Press, 2013.

Sherwood, Robert E. *Roosevelt and Hopkins An Intimate History*. New York: Harper & Brothers, 1948.

Stettinius, Edward R. Jr. *Lend-Lease Weapon for Victory*. New York: The Macmillan Company, 1944

Vigneras, Marcel. *Rearming the French*. Washington, DC: Office of the Chief of Military History, Department of the Army, 1957.

White, Gerald T. *Training of Foreign Nationals by the AAF 1939-1945*. Washington, DC: Air Historical Office, Headquarters, Army Air Forces, August 1947. (AAF

Historical Study No. 64).

Web Sites

49th Fighter Wing Public Affairs Office, "German Air Force," Jul 14, 2008, at http://www. holloman.af.mil/ library/factsheets/factsheet.asp?id=4364.

80th Training Wing Public Affairs Office, "Euro-NATO Joint Jet Pilot Training Program (ENJJPT)," Oct 17, 2012, at http://www.sheppard.af.mil/library/factsheetspage/ factsheet.asp?fsID=5168.

AETC/HO, "Major Changes in Undergraduate Pilot Training 1939-1998," at http://www.aetc. af.mil/Shared/media/document/AFD-07130-081.pdf, accessed May 22, 2013.

_____, "Specialized Undergraduate Pilot Training Pilot Production FY92-F"Y02, at http://www.aetc.af.mil/pilot/productprodction/Chart.htm, accessed May 22, 2013.

"Air Chief Marshal Arthur T. Harris," The Pathfinder Museum, RAF Wyton, at http://www.raf.

Mod.uk/rafbramptonwyton/aboutus/sirarthurharris.cfm, accessed Aug 2, 2013.

Air Force History and Museum Program (HQ AF/HO), "Overview," n. d., at http://www. airforcehistory.af.mil/overview/index.asp, accessed May 14, 2013.

"Air of Authority - A History of RAF Organisation," at http://www.rafweb.org/, accessed Jun 4, 2114.

"America and the Second World War," America in the World Multipoint Briefings, at http://americaintheworld.typepad.com/briefings/2008/08/america-and-t-1.html, accessed May 11, 2013.

Bate, Nicola J. "The Arnold Scheme 1941-1943," at http://www.arnold-scheme.org/, Mar 27, 2013.

"Charles Lindbergh An American Aviator," 1998-2007, at http://www.charleslindbergh.com/ americanfirst/speech3.asp, accessed May 11, 2013.

Commonwealth War Graves Commission, http://www.cwgc.org/find-war-dead.aspx?cpage= 1&sort= name&order=asc, accessed Jun 13, 2014.

"Defense Language Institute English Language Center," n. d., at http://www.dlielc.edu/, accessed Aug 15, 2013.

Department of State, Office of the Historian, "Neutrality Acts of 1930s," at http://history. state.gov/milestones/1921-1936/Neutrality_acts, accessed Jun 3, 2013.

"Eagle Squadrons" at http://fourthfightergroup.com/eagles/index.html, accessed May 24, 2013.

Hill Air Force Base Museum, "AT-6A Texan," http://www.hill.af.mil/library/factsheets/ factsheet.asp?id=5665, accessed Jun 9, 2013.

"Laos – A Secret Pilot War," Mar 12, 2011, at http://heritageflightgeardisplays.wordpress. com/2011/03/12/laos-a-secret-pilot-war/, accessed Aug 8, 2013.

Margerison, Ava, 2Lt, 82 TW/PA, "Italian instructor pilot reaches 5,000 hour milestone," Jan 15, 2014, at http://www.sheppard.af.mil/news/story.asp?id=123376853.

Martin, Jean, "The Golden Years: John Clark," Selma Times-Journal, Oct 16, 2010, at http://www.selmatimesjournal.com.2010/10/16/the-golden-years-john-clark/, accessed Jun 6, 2014.

National Museum of the United States Air Force, "Beech A-10 Wichita," http://www.nationalmuseum.af.mil/factsheets/factsheet. asp?id=483, accessed Jun 13, 2013.

_____, "Curtis AT-9 Jeep/Fledgling," at http://www.nationalmuseum.af.mil/factsheets/ factsheet.asp?id=482, accessed on Jun 15, 2013.

_____. "North American B-25," http://www.nationalmuseum.af.mil/factsheets/factsheet. asp?id=2509, accessed Jun 6, 2013.

_____, "Stearman PT-13D Kaydet," http://www. nationalmuseum.af.mil/factsheets/factsheet.asp?id=499, accessed Jun 9, 2013.

"Obituary John Clark," http://obits.al.com/obituaries/birmingham/obituary.aspx?pid=167605659, accessed Jun 6, 2014.

Sarota, David, "The Forgotten Math: Pre-WWII New Deal Saw Biggest Drop In Unemployment Rate in American History," Jan 6, 2009, at http://www.ourfuture.org/blog-entry\ 2009010206/forgotten-math-pre-wwii-new-deal-saw-fastest-drop-unemployment-rate-american-h, accessed May 11, 2013.

Thompson, William J. "America First Committee," Facts On File, Inc., at http://www. fofweb.com/ activelink2.asp?ItemID=WE52&iPin=EAHVIII008&SingleRecord=True, accessed May14, 2013.

Warbird Alley, "Cessna T-50 / UC-78 Bobcat," at http://www.warbirdalley.com/uc78.htm, accessed 15 Jun 2013.

_____ "Vultee BT-13 Valiant," http://www.warbirdalley.com/bt13.htm

"What the U. S. A. Thinks," *Life*, Jul 29, 1940, p. 20. 154, at Google Books, http://books. google.com/books?id=xz8EAAAAMBAJ &source=gbs_all_issues_r&cad=1, accessed May 11, 2013.

Newspapers

Montgomery *Advertiser*. Local newspaper for Montgomery, AL, Jun 1941 through Dec 1945. Microfilm in Alabama State Archives, Montgomery, AL.

Training News, bound volumes, 1941-1944, in Air University Directorate of History, Maxwell AFB. The *Training News* contained news about all the training schools assigned to the Southeast Air Corps Training Center/Eastern Flying Training Command. The newspaper underwent several names changes from 1940 to 1945: *Southeast Air Corps Training News*, Jul 1940 to Nov 22, 1942; *Army Air Forces Training Center News*, Nov 21, 1942 to Mar 26, 1943; *Training News* after Mar 27, 1943. The directorate has the papers for 1941 through 1944. For simplicity, I referred to the newspaper as the *Training News* throughout the manuscript.

Other

International Officer School History, n.d., provided by Mr. Gregory Baker, IOS.

Interview, author with Mr. Gregory L. Baker, Foreign Disclosure Officer, Air University, Maxwell AFB, AL, Aug 14, 2013.

Interview, author with Philip Taunton, Oakwood Cemetery Manager, Montgomery, AL, Aug 16, 2013.

E-mails as noted in the footnotes.

Sources of Photographs

1.1. New York World-Telegram and the Sun Newspaper Photograph Collection, Prints & Photographs Division, Library of Congress, Washington, DC, at http://www.loc.gov/pictures/item/2005680391/.

2.1. http://www.af.mil/AboutUs/Biographies/Display/ tabid/225/Article/107811/general-henry-h-arnold.aspx, accessed Aug 3, 2013.

2.2. EFTC History 7 Dec 1941–1 Jan 1943, Vol. 5, AFHRA Call # 222.01 7, Doc # 00146139.

2.3. Original photo from the Imperial War Museum.

2.4. Air Historical Branch, Royal Air Force, MOD/Crown Copyright.

3.1. Maxwell Field History, 20 Feb 1910–31 Dec 1938, AFHRA Call # 286.26, Doc # 00175337.

3.2. Gunter Field History 10 Aug 1940–30 Apr 1944, Vol. 3, AFHRA Call # 283.81–1, Doc # 00172952.

3.3. Organizational History Branch, Air Force Historical Research Agency (AFHRA), Maxwell AFB, AL.

3.4. AAF FTC History 1 Jan 1939–7 Dec 1941, Vol. 3, AFHRA Call # 220.01, Doc # 00145263.

3.5. AU/HO photo files.

3.6. AU/HO photo files.

3.7. AAF FTC History, 1 Jan 1939–7 Dec 1941, Vol. 3, AFHRA Call # 220.01, Doc # 00145263.

3.8. EFTC History 7 Dec 1941 - 1 Jan 1943, Vol. 5, AFHRA Call # 222.01 7, Doc # 00146139.

3.9. AU/HO photo files.

3.10. AFHRA Call # 283.81–103 Doc # 2–3725-16

3.11. AU/HO photo files.

3.12. Donated by Ms. Rose Koch, Clark daughter, to the AFHRA, Maxwell AFB.

3.13. Gunter Field History, 7 Dec 41 – Dec 1942, Vol. 3, AFHRA Call #283.81–2 Doc # 00172958

4.1. EFTC History Jan - Dec 1943, Vol. 6, AFHRA Call #222.01, Doc # 00146146.

4.2. EFTC History Jan - Dec 1943, Vol. 6, AFHRA Call #222.01, Doc # 00146146.

4.3. EFTC History Jan 1943 –Dec 1943, Vol. 5, AFHRA Call # 222.01 Doc # 00146145.

4.4. Gunter Field History, 1 Jan 1943 to 31 Jan 1944, Vol. 4, AFHRA Call # 283–81-3, Doc # 00172963.

4.5. Gunter Field History, 1 Jan 1943 to 31 Jan 1944, Vol. 4, AFHRA Call # 283–81-3, Doc # 00172963.

4.6. EFTC History Jan 1943 –Dec 1943, Vol. 6, AFHRA Call # 222.01, Doc # 00146146.

4.7. AU/HO photo files.

5.1. A1C Sondra Escutia, 49 FW/PA, "GAF FTC commemorates 55,555 flight hours," Oct 14, 2009, at http://www.holloman. af.mil/news/story.asp?id=123172680, accessed Jul 15, 2014.

5.2. Mr. William Clifton, 80 FTW History Office (80 TFW/HO), Sheppard AFB, Texas.

5.3. 2Lt. Ava Margerison, 82 TW/PA, "Italian instructor pilot reaches 5,000 hour milestone," Jan 15, 2014, at http://www.sheppard.af.mil/news/story.asp?id=123376853.

6.1. Thomas. H. Greer, "Individual Training of Flying Personnel," in Craven and Cate, Vol VI, p. 688.

6.2. Donated by Ms. Rose Koch, daughter of RAF First Lieutenant John Clark, to the AFHRA, Maxwell AFB, Alabama.

6.3. Donated by Ms. Rose Koch, daughter of RAF First Lieutenant John Clark, to the AFHRA, Maxwell AFB, Alabama

6.4. AU/HO photograph files

6.5. Donated by Ms. Rose Koch, daughter of RAF First Lieutenant John Clark, to the AFHRA, Maxwell AFB, Alabama

Notes

1. In addition to RAF and FFAF aircrew members, the AAF also trained aircrew cadets from Mexico, China, and other allied countries at Maxwell and Gunter Field during World War II. However, since the training programs for the RAF and the FFAF were two largest of these programs, this paper will focus only on the RAF and FFAF training programs. The US Army Air Corps (USAAF after June 19, 1941) established the Southeast Air Corps Training Center on July 8, 1940 and redesignated it as the Eastern Flying Training Command on July 31, 1943. The AAF inactivated the EFTC on December 15, 1945. Thomas A. Manning, *History of Air Education and Training Command, 1942–2002* (Randolph AFB, TX: Headquarters Air Education and Training Command Office of History and Research, Headquarters, 2005).
2. The AAF trained about 4,300 RAF flight crew members at Maxwell and Gunter Fields from June 1941 to early 1943 when the AAF and RAF terminated the program. Most of these went on to more advanced flying schools after graduation from the Maxwell and Gunter programs. The AAF trained about 6,000 RAF flight crewmen overall at various locations, including Maxwell and Gunter Fields, and another 6,000 received training at RAF-run contracted flight training schools in the United States, the British Flight Training Schools (BFTS), during the war. The Empire (British Commonwealth after June 1942) Air Training Program (BCATP), however, trained 131,000 during the war. Thus, the RAF pilots trained at Maxwell and Gunter Fields represented 36 percent of those trained by the AAF and the BFTSs together during the war, but the 12,000 some RAF flight crew trained in the United States were only about 9 percent of the total trained between the US training programs and the BCATP.
3. The AAF graduated a total of 1,561,288 from all phases and types of flight training from July 1939 to August 1945. Of these, 193,440 were pilots from the advanced school; 43,051 bombardiers and bombardier-navigators; 48,870 navigators; and 309,236 flexible gunners. "The US Army Air Forces at War," *Air Force Magazine* (Vol. 28, June 1995), p. 36. Ironically, despite this massive wartime training program, there are few complete studies of this massive training program that contributed so much to the Allied victory in World War II. The BCATP produced 131,000 aircrew members for the RAF, the Royal Canadian Air Force, the Royal Australian Air Force, the Royal New Zealand Air Force, and the South African Air Force during World War II. Appendix B, The BCATP Air Crew Graduates, Spencer Dunmore, *Wings For Victory* (Toronto: McClelland and Stewart, 1994), p. 361. In addition nearly 2,000 Free French, about 900 Czechoslovakians, 677 Norwegians, 450 Poles, and about the same number of Belgians and Dutch received training at BCATP schools. Peter C Conrad, *Training for Victory: The British Commonwealth Air Training Plan in the West* (Saskatoon: Western Producer Prairie Books, 1989), p. 13. After reviewing several of these books, several hundred Americans apparently received flight training at BCATP schools, but I could not find any specific number. At its peak level of operation in mid-1943, the BCATP had 107 schools and 184 ancillary units at 231 sites,

10,906 aircraft, and personnel numbering 104,113 men and women. By the time the RAF terminated the program in March 1945, it had cost of over $2 billion and had provided almost 50 percent of the total aircrew for British and Commonwealth flying operations during the war. F. .J. Hatch and Norman N. Hillmer, "British Commonwealth Air Training Plan," *The Canadian Encyclopedia*, at http://www thecanadianencyclopedia.com/articles/british-commonwealth-air-training-plan, accessed May 29, 2013. The paucity of secondary works on AAF flight training contrasts starkly with the secondary literature on the BCATP. Other works on the BCATP include Ted Barris, *Behind The Glory: The Plan that Won the Allied Air War* (Markham, Ontario: Thomas Allen & Son Publishers, 2000); Captain Dave Becker. *Yellow Wings: The Story of the Joint Air Training Scheme in World War 2* (Pretoria: The South African Air Force Museum, 1989); F. J. Hatch, *Aerodrome of Democracy: Canada and the British Commonwealth Air Training Plan 1939–1945* (Ottawa: Canadian Department of National Defence, 1983); John. McCarthy, *A Last Call of Empire: Australian Aircrew, Britain and the Empire Air Training Scheme* (Canberra: Australian War Memorial, 1988); MacDonald, J.F. *The War History of Southern Rhodesia 1939–1945* (Bulawayo, Zimbabwe: Authority of the Government of Southern Rhodesia, 1947); I. Norman Smith, *The British Commonwealth Air Training Plan* (Toronto: Macmillan Company of Canada Limited, 1941).

4. Because I currently reside in Montgomery, Alabama, and work as an Air Force historian for Air University, located at nearby Maxwell Air Force Base, I have access to the Muir S. Fairchild Research and Information Center (MSFRIC), the largest library in the US Air Force and the largest library of the US Government outside of Washington, DC. I developed this assessment of the limited secondary works on RAF training by the USAAF during World War II on the holdings of the MSFRIC. The MSFRIC has even fewer works on the French Air Force, limiting my review of secondary works in English on the history of the French Air Force, much less those in French, which I cannot read anyway. However, I suspect that there is probably even less on this aspect of the history of the FFAF during World War II as well, given the paucity of works in English on AAF training of RAF aircrew.
5. Thomas B. Buell, John N. Bradley, and John H. Bradley, *The Second World War: Europe and the Mediterranean* (Garden City, NY: Square One Publishers, 2002), p. 43.
6. AVM Grahame G. Dawson, Chief, Technical Branch, RAF, to John G. Winant, US Ambassador to Britain, "Enclosure B, Note on Scheme for Training Pilots and Observers in the U.S.A.," Mar 1941, Box 360, Papers of Harry L. Hopkins, Franklin D, Roosevelt Presidential Library, Hyde Park, NY. (Cited hereafter as Hopkins Papers)
7. Spencer Dunmore, *Wings For Victory* (Toronto: McClelland and Stewart, 1994), pp. 61–63.
8. David Sarota, "The Forgotten Math: Pre-WWII New Deal Saw Biggest Drop In Unemployment Rate in American History," Jan 6, 2009, at http://www.ourfuture.org/blog-entry\ 2009010206/forgotten-math-pre-wwii-new-deal-saw-fastest-drop-unemployment-rate-american-h, accessed May 11, 2013.
9. Department of State, Office of the Historian, "Neutrality Acts of 1930s," at http://history.state.gov /milestones/1921–1936/Neutrality_acts, accessed Jun 3, 2013. "America and the Second World War," *America in the World Multipoint Briefings*, at http://americaintheworld.typepad.com/briefings/2008/08/america-and-t-1.html, accessed May 11, 2013. "What the U. S. A. Thinks," *Life*, Jul 29, 1940, p. 20. 154, at Google Books, http://books. google.com/books?id= xz8EAAAAMBAJ &source=gbs_all_issues_r&cad=1, accessed May 11, 2013. See, for example, the speeches of Charles Lindberg, October 13, 1939 and August 4, 1940, at "Charles Lindbergh An American

Aviator," 1998–2007, at http://www.charleslindbergh.com/americanfirst/speech3.asp, accessed May 11, 2013.

10. Britain had begun World War II on September 3, 1939, with about $4.5 billion in dollars, gold, and investments in the United States. In the following 16 months, Britain had managed to produce almost $2 billion more from gold sales, exports, and other sources. However, in the same period, Britain had paid out nearly $4.5 billion for war materiel from the United States and other countries which demanded payment in currency. By December 1940, Britain only had about $2 billion left to purchase war materiel of which $1.5 billion was pledged to pay for war materiel ordered in the United States but not yet delivered. By mid-December 1940, the British government had practically ceased additional contracting. Edward R. Stettinius, Jr., *Lend-Lease Weapon for Victory* (New York: The Macmillan Company, 1944), pp. 57–61.
11. David L. Roll, *The Hopkins Touch: Harry Hopkins and the Forging of the Alliance to Defeat Hitler* (Oxford, UK: Oxford University Press, 2013), pp. 73–75. Lothian quoted in Roll, p. 73.
12. Ibid., pp. 76–77. "Fireside Chat" ("The Great Arsenal of Democracy"), Dec 29, 1940, Roosevelt Speeches, *The American Presidency Project*, at http://www.presidency.ucsb.edu/franklin_roosevelt.php, accesesd May 29, 2013 (Cited hereafter as Roosevelt Papers).
13. Roll, pp. 78–79.
14. Ibid., pp. 78–96. See Robert E. Sherwood, *Roosevelt and Hopkins: An Intimate History* (New York: Harper & Brothers, 1948), pp. 257–58, for Hopkins's recommendations for aid to Britain, based on his observations and meetings with British officials, submitted to President Roosevelt after Hopkins return to the United States.
15. "Give Us the Tools," Feb 9, 1941 Radio Broadcast, "Selected Speeches of Winston Churchill," Churchill Centre and Museum at the Churchill War Rooms, at http://www.winstonchurchill.org/learn/speeches/speeches-of-winston-churchill/97-give-us-the-tools, accessed May 13, 2013.
16. "Annual Message to Congress on the State of the Union, Jan 6, 1941, Roosevelt Papers.
17. William J. Thompson, "America First Committee," Facts On File, Inc., at http://www.fofweb.com/activelink2.asp?ItemID=WE52&iPin=EAHVIII008&SingleRecord=True, accessed May 14, 201_).
18. Roll, pp. 80, 90, 94, 99.
19. Jay E. Hines, *History of Foreign Training in ATC* [Air Training Command], *1941–1976* (Randolph AFB, TX: History and Research Division, Office of the Chief of Staff, Headquarters Air Training Command, [1977]), pp. 5–6.
20. Ibid.
21. "Lend Lease Act, 11 March 1941," Naval Historical Center, Washington, DC, at http://www.history. navy.mil/faqs/faq59–23.htm, accessed May 13, 2013.
22. Roll, pp. 100–101, 105. Sherwood, p. 278.
23. History of US air force organization from 1907 to World War II: Aeronautical Division, US Signal Corps, August 1, 1907; Aviation Section, US Signal Corps, July 18, 1914; Bureau of Aircraft Production and Division of Military Aeronautics, May 20, 1918; Army Air Service, May 24, 1918, Army Air Corps, July 2, 1926; Army Air Forces, June 20, 1941. The Army Air Corps remained as a combatant arm of the Army coequal with the Infantry, Armor, and Artillery, and the Army Air Forces had operational command of the fielded combat air forces. Air Force History and Mu-

seum Program (HQ AF/HO), "Overview," n. d., at http://www.airforcehistory.af.mil/overview/index.asp, accessed May 14, 2013.

24. Sherwood, pp. 258–59.

25. John W. Huston, Maj Gen, USAF Retired, ed., *American Airpower Comes of Age General Henry H. "Hap" Arnold's World War II Diaries* (Maxwell AFB, AL: Air University Press, 2002), Vol. I, pp. 129–31. This two-volume book is a combination of historical commentary by the editor and verbatim entries from Arnold's wartime diaries. Subsequent footnotes for this source will differentiate Huston's narrative/commentary from Arnold's diary entries by noting the latter with a date and an annotation that the information comes from a verbatim diary entry.

26. Clayton Knight was an American artist-illustrator who had served with the Royal Flying Corps and the RAF in 1917–18 during World War I and, later, along with Royal Naval Air Service veteran E. Homer Smith, formed a committe in early summer 1940 to recruit Americans to fly with the RCAF and the RAF. The Committee had discovered that many prospective American pilots had around 80 flying hours, and these schools could provide the additional 80 hours to meet a minimum RAF requirement of 150 hours at less costs and faster than recruiting pilots with little or no flying training. The Clayton Knight Committee sent 900 trained aircrew and 1,450 trainees to the RCAF and 300 pilots to the RAF before the attack on Pearl Harbor. A total of about 9,000 Americans went to Canada to join the RCAF by that time. After Pearl Harbor, 1,759 of them returned to the United States and joined one of the US Armed Forces. By the end of the war, an additional 2,000 transferred to the American military, and about 5,000 completed their wartime service with the RCAF. "The Americans in the RCAF," at http//www.bombarcommandmuseumrcaf.html, n. d., accessed Feb. 3, 2015. Quinn, pp. 35–39.

27. After completing their training at initial contract schools in the United States, these American volunteer pilots then travelled to Britain through Canada for assignment to the Eagle Squadrons. See "Eagle Squadrons" at http://fourthfightergroup.com/eagles/index.html, accessed May 24, 2013, for more information about the Eagle Squadrons.

28. Letter, Air Commodore George C. Pirie, Air Attaché, to Harry Hopkins, Mar 17, 1941, Box 360, of Hopkins Papers. AVM Grahame G. Dawson, Chief, Technical Branch, RAF, to John G. Winant, US Ambassador to Britain, "Enclosure B, Note on Scheme for Training Pilots and Observers in the U.S.A.," Mar 1941, Box 360, Hopkins Papers. Memorandum for Gen Arnold, "British Flying Schools," Apr 7, 1941, AFHRA Folder Call # 2207271–4 British 1941, Doc # 00145958. Phil Butler and Dan Hagedorn, *Air Arsenal North America Aircraft for the Allies 1938–1945 Purchases and Lend-Lease* (Hinckley, UK: Midland Publishing, 2004), p. 63.

29. Huston, pp. 130–35. H. H. Arnold, *Global Mission* (New York: Harper & Brothers, 1949), pp. 215–40.

30. Summary of April 9–May 1, 1941, entries, Arnold's diary, in Huston, pp.135–73. During his visit to Britain, Arnold also had time to get personal information about RAF strategic bombing efforts. RAF pilots insisted that precision bombing required at least 20 seconds of straight flying in approaching the target but was never used, except at high altitudes, because of German antiaircraft defenses and fighters. These pilots recommended the United States develop a bomber that could accurately bomb after 20 seconds straight flying. Arnold noted that the Air Corps should study such a proposal and that the RAF was not using the Sperry (Norden) bombsight. He wrote that the RAF had "much to learn about bombing but who am I to question two years of bombing in real war." April 12 and April 17 entries, Arnold's diary, in Huston, pp,

139, 147–48. In his memoirs, Arnold wrote, "That was my introduction to a general campaign that later developed into an official deprecation of our daylight bombing and a constantly nagging effort to get us to go along with the R.A.F. in their night bombing. Arnold, pp. 215–16, 225. *Author's Note*: Because of severe losses to German antiaircraft defenses in the RAF's initial attempts at daylight strategic bombing of Germany, the RAF went to nighttime bombing which led to area bombing with incendiary bombs, given the difficulty of finding, much less, hitting discrete targets at night. Also, the RAF did not have an accurate bombsight, and the United States refused to provide the RAF with the top-secret Norden bombsight. The irony was that, despite the detailed precautions to guard the highly classified Norden bombsight, the *Luftwaffe* had obtained the plans of the bombsight in 1937 when a German immigrant, working in the Sperry plant, copied the plans of the bombsight, hid the copied plans in his clothes during a trip back to Germany, and sold the plans to the Gestapo. Additionally, Arnold developed an admiration for the stoicism of the British people who maintained good morale and resolve, despite the regular bombing of their cities, causing such death and destruction. Yet, he steadfastly maintained that strategic bombing could not only destroy German (and later Japanese) industrial production but also severely degrade civilian morale.

31. April 13, 1941 entry, Arnold's diary, in Huston, pp. 140–41. Arnold, p. 216.
32. Arnold, p. 216.
33. Quoted in Footnote 68, Huston, p. 187.
34. Called variously the Arnold Scheme, the Arnold Plan, or the Arnold Program, See web site "The Arnold Scheme 1941–1943," currently under construction as "The Official Website of "The Arnold Scheme (1941–1943) Register," http://www.arnold-scheme.org/, last updated, Mar 27, 2013, by Nicola J Bate (cited hereafter as The Arnold Scheme).
35. Huston, p. 179.
36. Arnold, p. 245.
37. Letter, Robert A. Lovett to Harry L. Hopkins, Apr 29, 1941; Memorandum, Oscar Cox to Harry Hopkins, "Training of Pilots for PBY-Flying Boats," n.d.; Letter, Harry Hopkins to Robert A. Lovett, May 1, 1941; and Memorandum, John J. McCloy to Harry Hopkins, May 20, 1941; Hopkins Papers.
38. Eugene C. Gerhart, *America's Advocate: Robert H. Jackson* (Indianapolis: The Bobbs-Merrill, Inc., 1958), pp. pp. 223–27.
39. Gerald T. White, *Training of Foreign Nationals by the AAF 1939–1945* (Washington, DC: Air Historical Office, Headquarters, Army Air Forces, August 1947) (AAF Historical Study No. 64), pp. 4–5.
40. *History of the AAF Flying Training Command and Its Predecessors, 1 January 1939 to 7 July 1943* (cited hereafter as *AAF FTC History)*, Vol. 1, p. 176, AFHRA Call # 221.01 1 Jan 1939–7 Jul 1943 V. 1, Doc # 00146039.
41. Edward R. Stettinius, Jr., *Report to the 78th Congress on Lend-Lease Operations from the Passage of the Act, March 11, 1941, to December 31, 1942*, Jan 25, 1943, pp. 21–22
42. In addition to these three programs conducted by or overseen by the AAF, US Navy Vice Admiral John H. Towers, Chief of the Bureau of Aeronautics, created a fourth program to train RAF and Fleet Air Arm cadets and the crews for PBY Catalina flying boats that Britain obtained with Lend-Lease funds. Towers approached the President about this idea in early 1941, and, in early July 1941, RAF and Royal Navy flight cadets left Britain for the first course at Naval Air Station Pensacola, Florida.

They underwent flight training alongside US Navy and US Marine flight cadets. By the time the program was terminated, some 4,000 British aircrew (2,500 pilots, 545 navigators, about 600 radio operators, and a small number of helicopter pilots and maintainers) had graduated from the US Navy pilot training program. John Golley, *Aircrew Unlimited The Commonwealth Air Training Plan during World War 2* (Sparkford, UK: Patrick Stephens Limited, 1993), pp. 91–123. Butler and Hagedorn, p. 64. Guinn, p. 24.

43. Hines, p. 10.
44. Memorandum, Henry L. Stimson to War Department Budget Officer, "Establishment of British Air Schools in the United States – Lend-Lease Act – Requisitions Nos, 88 and 89 (British); Hopkins Papers. Requisition 88 details the specifications for six contractor schools, and Requisition 89 details the specifications for the refresher course. Memorandum for Gen Brett, author unknown, "Report of Conference in General Johnson's Office April 30th," Apr 30, 1941, AFHRA Folder Call # 220.7272–4 1941 British 1941, Doc # 00145958. Hines, pp. 8–9.
45. Hines, pp. 8–9.
46. Memo for Maj Gen (George) Brett, Assistant to Gen Arnold, "Report of Conference Held in General Johnston's Office April 30th," Apr 30, 1941, AFHRA Folder Call # 2207271–4 British 1941, Doc # 00145958, AFHRA. Memo for Maj Gen (George) Brett, Assistant to Gen Arnold, "Training of British Student Pilots Under the Six School Plan," May 19, 1941, AFHRA Folder Call # 2207271–4 British 1941, Doc # 00145958. Hines, p. 14. In addition to training RAF pilots, the BFTS also graduated 558 American pilots as part of an "exchange" training program between November 17, 1942 and June 18, 1944, when the last group of Americans graduated. Hines, pp. 12, 14.
47. Memo, Lt Col L. W. Miller, Chief of Fiscal Division, to Gen Arnold, "Request for Allocation of Lend-Lease Funds," May 29, 1941, AFHRA Folder Call # 2207271–4 British 1941, Doc # 00145958.
48. Hines, pp. 10–11.
49. Jerome A. Ennels and Wesley Phillips Newton, *The Wisdom of Eagles A History of Maxwell Air Force Base* (Montgomery, AL: Black Belt Press, 1997), p. 82.
50. *History of the Army Air Forces Eastern Flying Training Command, First Instalment, 1 January 1939-7 December 1941* (cited hereafter as *AAF EFTC History*), Vol. 1, pp. 1-6, 562-63, AFHRA Call # 222.01 1 Jan 1939-7 Dec 1941 v. 1, Doc # 001416132. "Britons To Reach Alabama Sunday," *Montgomery Advertiser*, Jun 7, 1941, p. 7.
51. "U.S. and British Cadets to March," *Montgomery Advertiser*, Oct 19, 1941, p. 2.
52. Air University Directorate of History (AU/HO), *Cradle of Airpower Education A Short History of The Air University, Maxwell Air Force, and the 42nd Air Base Wing* (Maxwell AFB, AL: Air University Directorate of History, June 2012), pp. 3-4.
53. *AAF EFTC History, 1 Jan 1939–7 Dec 1941*, vol. 1, pp. 1–6, AFHRA Call # 222.01 1 Jan 1939–7 Dec 1941 v. 1, Doc # 001416132. The histories of the Eastern Flying Training Command, Maxwell Field, and Gunter Fields indicate that the RAF (and later the Free French) flight cadets used the same curricula under the same conditions as RAF cadets at SEATCTC's civilian contractor schools. These histories cover the revisions in the training programs as the war progressed, and the AAF FTC incorporated changes as overall training progressed. See Appendix 6 for the AAF training program in 1943.
54. *Station History Army Air Force Pilot School (Basic) Gunter Field Alabama, First Instalment, Activation to 7 December 1941*), pp. 8–9, 12–17, 34, 62. (Air University

Directorate of History) (Cited hereafter as *Gunter Field History*) Author's Note: The city of Montgomery had opened the original municipal airport in 1929. By 1940, it consisted of 187 acres of a 600-acre tract between the Seaboard Airline Railroad and Alabama State Highway 14 across from Kilby Prison. The AAC precedent was to name airfields after deceased airmen. However, after Mayor Gunter died on December 4, 1940, city officials petitioned the AAC to name the airfield after Gunter, and the AAC reluctantly agreed. With the transfer of this airport to the Army Air Corps, Eastern Airlines temporarily operated from Maxwell Field until the city constructed a new airport southwest of the city. On July 1, 1943, city officials dedicated the new airport in honor of Ensign Clarence Moore Dannelly Jr., a US Navy pilot killed in a 1940 training accident and considered as Montgomery's first casualty of World War II. During the war the new municipal airport also served as Gunter Field Auxiliary Field #6. In late Nov 1945, the AAF declared Auxiliary Field #6 as excess, and the airfield returned to civilian use as the Montgomery municipal airport. "Dannelly Field Dedication Set For Thursday, Montgomery *Advertiser*, Jun 27, 1943, pp. 1, 2. "Mail Planes Land Again as Dannelly Field Opens," Montgomery *Advertiser*, Jul 2, 1943, pp. 1, 7. "City Reclaims Use as Airport," Montgomery *Advertiser*, Nov 21, 1945, p. 3. Quinn, pp. 304–5, 313. See the *Gunter Field History* above and Guinn, pp. 304–11, for details of the acquisition and development of Gunter Field and its basic flying school from summer 1940 through summer 1941 and the arrival of the first class of RAF flight cadets.

55. *AAF EFTC History, 1 Jan 1939–7 Dec 1941*, Vol. 1, p. 557, AFHRA Call # 221.01 1 Jan 1939–7 Jul 1941 v. 1, Doc # 00146032. "British Airmen's Flying Here Scheduled To Start Saturday," Montgomery *Advertiser*, Jun 3, 2942, pp. 1, 5.
56. *AAF EFTC History, 1 Jan 1939–7 Dec 1941*, Vol. 1, p. 557, AFHRA Call # 221.01 1 Jan 1939–7 Jul 1941 v. 1, Doc # 00146032, pp. 173–74. Quinn, pp. 44–46, 186–87. Also in July 1941, the Royal Navy established the British Air Delegation, also located in Washington, DC, to administer and oversee the training of RAF and Fleet Air Arm aircrew (flying boat, carrier fighter, and torpedo and dive bomber pilots; navigators; and radio operators/gunners) by the US Navy (Towers Scheme). British students for these training programs also arrived at Halifax by ship and processed at the No. 1 PD, Toronto, and later No. 32, Moncton, to their schools in the United States. Quinn, p. 186.
57. "RAF Officers of High Rank Coming Here," Montgomery *Advertiser*, Oct 9, 1941, p. 7. "Air Chief Marshal Arthur T. Harris," The Pathfinder Museum, RAF Wyton, at http://www.raf.mod.uk/rafbramptonwyton/a boutus/sirarthurharris.cfm, accessed Aug 2, 2013. By the end of World War II, Air Marshal Harris had received the nickname "Bomber Harris" for advocating the area bombing of German cities with incendiaries in an attempt to break the will of the German people.
58. Quinn, pp. 44–59.
59. "Second Contingent Of British Youths Train In SEACTC," *Training News*, Jul 26, 1941, p. 1. *AAF EFTC History, 1 Jan 1939–7 Dec 1941*, Vol. 2, pp. 563–64, AFHRA Call # 222.01 AAF EFTC 1 Jan 1939–7 Dec 1941 v. 2, Doc # 0014033. In late May/early June 1941, WC Hogan has toured the AAF flying schools in Alabama, Georgia, and South Carolina and was embarrassed by the large number of ill-fitting khaki uniforms, worn by the newly arrived RAF cadets. He then arranged with RAF liaison officers at the schools to pay for alterations to these uniforms. Apparently, the Americans at the schools had similar thoughts, and, soon afterwards, the Army Air Corps began issuing fitted American khaki uniforms and summer flying gear to the RAF students. Quinn, p. 65. Quinn called it "acclimatization training." Quinn, pp. 205–7.

60. "Maxwell Field Set To Receive 663 New Cadets," Montgomery *Advertiser*, Oct 4, 1941, p. 7. "2 Trains Due Monday with 750 Britishers," Montgomery *Advertiser*, Oct 12, 1941, p. 13.
61. "Air of Authority - A History of RAF Organisation," at http://www.rafweb.org/, accessed Jun 4, 2114.
62. White, p. 12. *AAF EFTC History, 1 Jan 1939–7 Dec 1941,* Vol. 1, pp. 566–67, 73–74, AFHRA Call # 222.01 1 Jan 1939–7 Dec 1941 v. 1, Doc # 001416132. Craven and Cate, p. 558. The RTU program, like phases of the AAF training program, underwent changes in length and content, which the EFTC histories document. The RAF cadets for RTC Classes 42-E, 42-F, and 42-G completed the school and graduated with 637, 1,112, and 1,346 American cadets, respectively. Apparently, the Americans and the RAF cadets attended separate sections as the school had separate curricula with the one for the RAF cadets as depicted in Table III-1. *A History of Maxwell Field 7 December 1941 – 31 December 1942* (Cited hereafter as *Maxwell Field History*),Vol. 1, pp. 131, 150–52, AFHRA Call Number 286.01 7 Dec 1941–31 Dec 42, Vol. 1, Doc # 0015352. "New Center Opens Today," *Training News*, Sep 6, 1941, pp. 1, 8. In February 1941, the War Department had established RTCs at Kelly and Ellington Fields, Texas, and Santa Ana Army Air Base, California, to provide classification and preflight instruction for American pilot, bombardier, and navigator candidates. In April 1942, the AAF renamed them "prefight schools."
63. *Gunter Field History, Activation to 7 December 1941*, pp. 72–78, 99–100, 104–13. *Gunter Field History, 7 December 1941 to January 1943*, pp. 126–27. Classes 42-C. 42-D, and 42-D were completely RAF cadets. Other noted classes were mixed with a majority of RAF students.
64. *Maxwell Field History 7 Dec 1941 – 31 Dec1942*, Vol. 1, pp. 90–97, AFHRA Call # 286.26–2 7 Dec 1941 – 31 Dec 1942 v. 1, Doc # 00175352. "Flight School at Maxwell To Be Moved," Montgomery *Advertiser*, Jul 11, 1941, p.1.
65. Arthur R. Kooker, "Procurement at Flood Tide," in Craven and Cate, pp. 510-13.
66. *AAF EFTC History 7 Dec 41-1 Jan 43*, Vol. 2, pp. 1097-1101, 1115, AFHRA Call # 222.01 7 Dec 1941-1 Jan 1943 v. 2, Doc # 00146136. *Maxwell Field History 7 Dec 1941 – 31 Dec 1942*, Vol. 1, pp. 97- 98, AFHRA Call # 286.26-2 7 Dec 1941-31 Dec 1942 v. 1, Doc # 00175352.
67. Ennels and Newton, p. 90.
68. Ennels and Newton, p. 90.
69. Hewes quoted in Golley, p. 132. *Author's Note:* "Patter" is the ability of an instructor pilot to clearly and succinctly describe everything that he is doing while flying at a high level of accuracy to "show" the student the correct technique. Thus, apparently, Hewes was commenting on the lack of commentary and explanation from the instructor pilot who supposedly was teaching him how to be instructor pilot.
70. Golley, pp. 132-33.
71. *AAF EFTC History, 7 Dec 1941-1 Jan 1943*, Vol. 2, p. 1111, AFHRA Call # 222.01 7 Dec 1941-1 Jan 1943 v. 2, Doc # 00146136. The basic schools had 185 RAF instructors. The advanced two-engine schools had 45. Letter, Maj Gen Ralph Royce, Commander SEACTC, to Commanding General, AAF Flying Training Command, "United Kingdom Instructors," Oct 14, 1942, AFHRA Folder Call # 220.7271-6 1942-1944, Doc # 00145960. The advanced single-engine schools had 156 RAF instructors, and the CIS at Maxwell Field at 71 in training. *AAF EFTC History, 7 Dec 1941-1 Jan 1943*, Vol. 2, pp. 1111-12, AFHRA Call # 222.01, 7 Dec 1941-1 Jan 1943 v. 2, Doc # 00146136.

72. *AAF EFTC History, 1 Jan 1939-7 Dec 1941*, Vol. 2, pp. 558-61, AFHRA Call # 222.01 AAF EFTC 1 Jan 1939-7 Dec 1941 v. 2, Call # 00146132. Guinn, pp. 22, 44-46, 448-49. Guinn also lists the initial RAF liaison officers at the other SEATC schools in summer 1941.
73. *AAF EFTC History*, 7 Dec 1941-1 Jan 1943, Vol. 2, p. 1112, AFHRA Call # 221.01 7 Dec 1941-1 Jan 1943 v. 2, Call # 00146136.
74. AAF FTC, Vol. VII, p. 2093, AFHRA Call # 221.01 AAF FTC 1 Jan 1939-7 Jul 1943 v. 7, Call # 00146044.
75. Memo, author unknown, "Procedures for the Commissioning of United Kingdom Students and Disposition at the Completion of Pilot Training Course," n.d.; Letter, R. F. Heppell, to Cmdr, SEACTC, "Commissioning of R.A.F. Airmen," Jan 7, 1942; Letter, Capt T. A. Hancock, Acting Asst, Adjutant General, SEACTC, to all Commanding Officers, SEACTC Advanced Flying Schools, "Diplomas," Jan 28, 1942; AFHRA Folder Call # 286.26-2 v. 5 1941-1942, Doc # 00175366.
76. Memorandum, Brig Gen D. J., Chief, T and O Division, to Gen Brett, "Training of British Cadets," Jul 24, 1941, AFHRA Folder Call # 2207271–4 British 1941, Doc # 00145958. " 'Language of Americans Perturbs British Officer," *Training News*, Nov 15, 1941, p. 8.
77. *Maxwell Field History 7 Dec 1941 – 31 Dec 1942*, pp. 229–31, AFHRA Call # 286.26–2 7 Dec 1941 – 31 Dec 1942 v. 1 Doc # 00175352. Golley, pp. 129–30, 134–37. Guinn, pp. 314–15.
78. *AAF EFTC History, 1 Jan 1939–7 Dec 1941*, Vol. 2, pp. 563–64, AFHRA Call # 222.01 1 Jan 1939–7 Dec 1941 v. 2, Doc # 0014033. *AAF EFTC History, 7 Dec 1941–1 Jan 1943*, Vol. 2, pp. 1121–22, AFHRA Call # 222.01 7 Dec 1941–1 Jan 1943 v. 2, Doc # 0014036. This history reports similar problems at other AAF flight schools, assigned to SEACTC. Quinn, pp. 212–15.
79. "British Cadets Lose Whiskers But Not Without Lamentations," Montgomery *Advertiser*, Aug 21, 1941, p. 24. "Gen. Weaver Here On Bearded Britons," Montgomery *Advertiser*, Oct 24, 1941, p. 8.
80. *AAF EFTC History, 1 Jan 1939–7 Dec 1941*, Vol. 2, pp. 563–64, AFHRA Call # 222.01 1 Jan 1939–7 Dec 1941 v. 2, Doc # 0014033. *AAF EFTC History, 7 Dec 1941–1 Jan 1943*, Vol. 2, pp. 1121–22, AFHRA Call # 222.01 7 Dec 1941–1 Jan 1943 v. 2, Doc # 0014036.
81. *AAF EFTC History, 1 Jan 1939–7 Dec 1941*, Vol. 2, p. 565, AFHRA Call # 222.01 1 Jan 1939–7 Dec 1941 v. 2, Doc # 0014033.
82. See Appendix 6, which provided the numbers of cadets entering each class and each phase of the training program, the numbers of graduates from each of phase of the training program, eliminations, and deaths and the percentage of graduates from each class and phase.
83. *AAF EFTC History, 7 Dec 1941–1 Jan 1943*, Vol. 2, p. 1103, AFHRA Call # 222.01 7 Dec 1941–1 Jan 1943 v. 2, Doc # 0014036.
84. Hewes quoted in Golley, p. 130.
85. Quinn, pp. 191-93.
86. Letter, author unknown, to GC D. V. Carnegie, Aug 11, 1941 w/attached letter, author unknown, "Pilot Training of United Kingdom Personnel, Aug 11, 1941, AFHRA Call # 2207271-4 British 1941, Doc # 00145958. AAF EFTC History, 1 Jan 1939-7 Dec 1941,Vol 2, pp. 572-73, AFHRA Call # 222.01 AAF EFTC 1 Jan 1939-7 Dec 1941 v. 2, Doc # 0014033.

87. *AAF EFTC History, 1 Jan 1939-7 Dec 1941*, Vol. 2, pp. 571-72, AFHRA Call # 222.01 AAF EFTC 1 Jan 1939-7 Dec 1941 v. 2, Doc # 0014033.
88. *AAF EFTC History, 7 Dec 1941-1 Jan 1943*, pp. 1108-09, AFHRA Call # 222.01 AAF EFTC 7 Dec 1941-1 Jan 1943, v. 2, Doc # 0014036. AAF FTC, Vol. VIII, p. 2101, AFHRA Call # 221.01 1 Jan 1939-7 Jul 1943, v. 7, Doc # 00146044. Letter, Author Unknown, to SEACTC/G-3, "Comparison of British and American Accidents Reported for the Period 3/30/42 – 5/1/42, SEAAFTC," May 21, 1942, Ch XIX Tab, AAF EFTC History, 7 Dec 1941-1 Jan 1943, Vol. 3, AFHRA Call # 222.01 7 Dec 1941-1 Jan 1943, v. 3, Doc #00146137.
89. *AAF EFTC History, 7 Dec 1941-1 Jan 1943*, pp. 1103-04, AFHRA Call # 222.01 AAF EFTC 7 Dec 1941-1 Jan 1943 v. 2, Doc # 00140363. AAF EFTC History, 1 Jan 1943-31 Dec 1943, Vol. 2, p. 955 , AFHRA Call # 222.01 AAF EFTC 1 Jan 143-31 Dec 1943 v. 2, Doc # 0014142. *Gunter Field History 7 December 1941 to 1 January 1943*, p. 64.
90. *AAF EFTC History, 7 Dec 1941-1 Jan 1943*, Vol. 2, pp. 1106-08, AFHRA Call # 222.01 AAF EFTC 7 Dec 1941-1 Jan 1943 v. 2, Doc # 0014136. *Gunter Field History 7 Dec 1941 to Jan 1943*, pp. 65-71. Quinn, pp. 322-25.
91. *AAF EFTC History, 7 Dec 1941-1 Jan 1943*, p. 1104, AFHRA Call # 222.01 AAF EFTC 7 Dec 1941-1 Jan 1943 v. 2, Doc # 0014136.
92. Ennels and Newton, p. 88.
93. Ibid, p. 88. "Military Funeral Held For British Cadets," Montgomery *Advertiser*, Nov 14, 1941, p. 6.
94. *Gunter Field History 7 Dec 1941 to Jan 1943*, p. 310. Oakwood Cemetery is just north of the current Montgomery Police Department on N. Ripley Street. The graves of the 78 RAF students killed in training accidents at SEACTC schools between June 1941 and February 1943 are near the Hank Williams Memorial. The SEACTC placed a wooden cross on the graves of the first RAF cadets killed but, by October 1943, had replaced them with granite headstones. Since 1981, the Canadian Agency, Commonwealth War Graves Commission, an agency of the British government, has contracted with a private company in Birmingham, AL, to maintain the gravesites. The RAF officers, attending the US Air Force officer courses at Maxwell AFB, hold special ceremonies each Veteran's Day at the gravesite in memory of their fallen World War II comrades. Interview by author with Philip Taunton, Oakwood Cemetery Manager, Montgomery, AL, Aug 16, 2013. Letter, Squadron Leader G. B. Cuthburtson, Personnel Staff, RAF Delegation (?), to Group Captain C. E., Maude, "Visit to Oakwood Cemetery, Montgomery," Nov 24, 1942; Letter, A. Lewis Watson, Asst Secretary-General, Imperial War Graves Commission, Ottawa Ontario, Canada, to RAF Delegation, Washington, DC, "Oakwood Cemetery, Montgomery, Ala , U.S.A Photographs of graves," Oct 1, 1943; AU/HO Document Files, Maxwell AFB, AL. WC Hogan had intended that all RAF cadets killed during training would be buried in the RAF section of Oakwood Cemetery. However, he had not counted on the generosity of Americans as American private citizens, churches, and veterans' organizations in towns near several SEACTC flight schools training RAF cadets donated land for this purpose. By Nov 1941 with the internment of the first two Maxwell cadets, at least six cadets had been buried in these local cemeteries. Since all 78 RAF cadets were interned in the Oakwood Cemetery by Feb 1942 and the end of the Arnold Program, apparently the SEACTC had the bodies of these deceased cadets reinterred in Oakwood Cemetery. Quinn, p. 453. See the Commonwealth War Graves Commission web site, http://www.cwgc.org/find-war-dead.aspx?cpage=1&sort= name&order=asc,

for a list of the RAF flight cadets interred in Oakwood Cemetery.

95. Letter, Mrs. Olive Shotbolt, Apr 29, 1942, in *Gunter Field History 7 Dec 1941 to Jan 1943*, p. 315.

96. Letter, Mrs. Mary E. Clare, n. d., in *Gunter Field History, 7 Dec 1941 to Jan 1943*, p. 316.

97. See references to the *Montgomery Advertiser* before December 1941.

98. *AAF EFTC History, 1 Jan 1939–7 Dec 1941*, Vol. 2, pp. 563–64, AFHRA Call # 222.01 AAF EFTC 1 Jan 1939–7 Dec 1941 v. 2, Doc # 0014033. *AAF EFTC History, 7 Dec 1941–1 Jan 1943*, Vol. 2, p. 1129, AFHRA Call # 222.01 AAF EFTC 7 Dec 1941–1 Jan 1943 v. 2, Doc # 0014136. Quinn, pp. 317, 321.

99. "Huntingdon College Club To Honor British Club," Montgomery *Advertiser*, Oct 19, 1941, Section Two, p. 3. "50 Cadets Invited To Lowndesboro," Montgomery *Advertiser*, Sep 21, 1941, p. 10. E-mail, Angela Wheeler, Library Director, Hayneville/Lowndes County Library, to author, "Debardeleben Plantation in Lowndesboro," Aug 8, 2013.

100. Quinn, p. 211. Hogan had persuaded his RAF superiors in Washington, DC, to allow his wife Venetia to accompany him to Maxwell Field during his assignment there. She became a great hit among the American military wives at Maxwell and Gunter Fields and proved to be a great "diplomatic" asset among Montgomery society. "Through her genuine friendliness, her teas and Maxwell Field club activities with RAF and USSAF wives Montgomery country clubs, Venetia Hogan proved to be an able and conscientious of Great Britain. She undoubtedly removed great pressures from the shoulders of her husband, who was already overloaded with responsibilities and labor." Quinn, pp. 448–49, 453.

101. "R.A.F. Cadets Will March In Public Review," Montgomery *Advertiser*, Sep 20, 1941, p. 5. "R.A.F. Cadets Will March In Parade," Montgomery *Advertiser*, Oct 2, 1941, p. 7. "U.S. and British to March," Montgomery *Advertiser*, Oct 19, 1941, p. 2. "Armistice Day Program Will Accent Future," Montgomery *Advertiser*, Nov 7, 1941, p. 6. "Steel To Flash in Parade On Armistice Day," Montgomery *Advertiser*, Nov 9, 1941, p. 5. "Huge Parade Moves Down Dexter Today," Montgomery *Advertiser*, Nov 11, 1941, p. 1. Quinn, p, 209.

102. "RAF Cadets Plan British Style Dance," Montgomery *Advertiser*, Nov 18, 1941, p. 3. "Royal Air Force Pilots Trained at Maxwell," Montgomery *Advertiser*, Sep 12, 1993, p. 3G.

103. "Royal Air Force Pilots Trained at Maxwell," Montgomery *Advertiser*, Sep 12, 1993, p. 3G.

104. "Weaver Called 'Godfather' of R.A.F. of South," Montgomery *Advertiser*, Oct 14, 1941, p. 9.

105. *Maxwell Field History 7 Dec 1941 – 31 Dec 1942*, pp. 229–31, AFHRA Call # 286.26–2 7 Dec 1941 – 31 Dec 1942 v. 1 Doc # 00175352. "British Cadets Find It's Still Summer Here," Montgomery *Advertiser*, Dec 16, 1941, p. 13.

106. Billy Spinks, "700 British Cadets Throng City On Their First Liberty," Montgomery *Advertiser*, Oct 2, 1941, p. 7. "U.S. and British to March," Montgomery *Advertiser*, Oct 25, 1941, p. 1.

107. Evans quoted in Golley, p. 137. Guinn provides a list of "many" RAF flight cadets who married American women during their time in the United. While it appears to be complete, he does not directly said so. Quinn, pp. 547–50.

108. *AAF EFTC History, 1 Jan 1939–7 Dec 1941*, Vol. 2, pp. 563–64, AFHRA Call #

222.01 AAF EFTC 1 Jan 1939–7 Dec 1941 v. 2, Doc # 0014033. *AAF EFTC History, 7 Dec 1941–1 Jan 1943*, p. 1129.

109. "Miss Wells and Mr. Trotter Wed at St. John's Church," Montgomery *Advertiser*, Mar 7, 1942, p. 3. "Royal Air Force Pilots Trained at Maxwell," Montgomery *Advertiser*, Sep 12, 1993, p. 3G. See also, "Author's Note: John Clark" in Appendix.
110. Quinn, pp. 450–52, 454.
111. "British To Shift U.S. RAF Training," *Training News*, Aug 1, 1942, p. 1.
112. Table 10, Appendix C, "Output of Pilots and Other Aircrew—Dominion Sources," Golley, p. 179.
113. Letter, Col L. S. Smith, Director of Individual Training, to Commander, AAF Flying Training Command, "Curtailment of British Flying Training in the United States," Jul 3, 1942, AFHRA Folder Call # 220.7271–5, 353 British 1942, Doc # 00145959.
114. "No Doubt About Victory, Avers R.A.F. Marshal," Montgomery *Advertiser*, Oct 15, 1941, p. 1.
115. "British Return Flags As Cadets Finish Training," *Training News*, Mar 6, 1943, pp. 1, 3. "RAF Returns Two Flags, Taking Leave Of Southeast," Montgomery *Advertiser*, Feb 26, 1943, pp. 1, 3.
116. See Appendix 6 for detailed statistics of the RAF flight training program Jane 1941 to March 1943. The RAF ceased flight training on September 8, 1945 and closed its last two contractor school three days later. All RAF personnel left the United States by the end of September. Hines, p. 35.
117. Portal quoted in *AAF FTC History*, Vol. VII, p. 577, AFHRA Call #221.01 1 Jan 1939–7 Jul 1943, v. 7, Doc # 00146045.
118. MFR, W. F. K,. AAF HQ Flying Training Command, "British Training in the United States," Jun 29. 1942, AFHRA Folder Call # 220.7271–5, 353 British 1942, Doc # 00145959.
119. *AAF FTC History, 1 Jan 1939–7 Dec 1941*, Vol. VII, p. 577, AFHRA Call #221.01 1 Jan 1939–7 Jul 1943, v. 7, Doc # 00146045.
120. Michael Vigneras, *Rearming the French* (Washington, DC: Office of the Chief of Military History, Department of the Army, 1957), pp. 7-9, 10-11, 214-15.
121. Ibid., pp. 12-21.
122. White, p. 34.
123. Vigneras, pp. 198, 203.
124. Charles Christienne and Pierre Lossarague, *A History of French Military Aviation* (Washington, DC: Smithsonian Institution Press, 1986), p. 396.
125. White, pp. 34-35.
126. Vigneras, pp. 196-98.
127. *AAF FTC History*, Vol. 4, pp. 1005-1006, AFHRC Call # 222.01 1 Jan 1939-7 Jul 1943 v. 4, Doc # 00146142.
128. Requisition for Defense Articles, No. Fr. 10,122, Apr 23, 1943, AFHRA Folder 353 French (1943), Call # 220.7272-2 V. 1, 1943, Doc # 00145964.
129. Letter, Col D. D. Fitzgerald, Acting Chief of Staff, "French Aircrew Training," May 8, 1943, AFHRA Call # 220.7272-2 V. 1, 1943, Doc # 00145964. Letter, Col J. H. Mills, SEACTC Adjutant General, to HQ AAF Asst Chief of Staff, Training, "French Aircrew Training," May 27, 1943, AFHRA Folder Call # 220.7272-2 V. 1, 1943, Doc # 00145964. AAF Training Command approved the establishment of one pool of French students, located at Craig Field, Alabama, for new arrivals and eliminees,

and the Craig post commander would house, feed, process, and orient them. Letter, Col L. S. Moseley, Asst to Chief of Staff, A-3, HQ AAF, to Commanding General. AAF Training Command, "Aircrew Training for Frenchmen," Oct 25, 1943, AFHRA Folder Call # 220.7272-2 V. 3 1944 Doc # 00145966. *AAF EFTC 1 Jan 1943 – 31 Dec 1943*, p. 1014, Call # 222.01, Doc # 00146142.

130. *Gunter Field History, 1 Jan 1943 to 31 Jan 1944*, p. 121.
131. Letter, Capt. R. L. Holcomb, SEACTC Acting Asst Adjutant General, to Commander SEACTC, "French Aircrew Training," Jun 28, 1943, AFHRA Folder 353 French (1943), Call # 220.7272-2 V. 1, 1943, Doc # 00145964.
132. Letter, Lt Col James W. Porter, SEACTC Asst Adjutant General, to Commander, SEACTC, "French Training Program," Jul 10, 1943, AFHRA Folder 353 French (1943), Call # 220.7272-2 V. 1, 1943, Doc # 00145964.
133. Letter, Col D. D. Fitzgerald, Acting Chief of Staff, "French Aircrew Training," May 8, 1943, AFHRA Folder Call # 220.7272-2 V. 1, 1943, Doc # 00145964.
134. *AAF EFTC History, 1 Jan 1943 – 31 Dec 1943*, Vol. 2, pp. 1011-1012, AFHRA Call # 222.01 AAF EFTC 1 Jan 1943-31 Dec 1943 v. 2, Doc # 00146142. Requisition for Defense Articles, No. Fr. 10,122-A, Jun 8, 1943; Letter, Capt. R. L. Holcomb, SEACTC Acting Asst Adjutant General, to Commander SEACTC, "French Aircrew Training," Jun 28, 1943, AFHRA Folder 353 French (1943), Call # 220.7272-2 V. 1, 1943, Doc # 00145964.
135. "French Cadets At Tuscaloosa Find Life Here 'Le Paradis'," *Training News*, Aug 14, 1943, pp. 1, 7. "First Contingent of 'Escadrille Francais' Arrives At Turner For Advanced Training," *Training News*, Nov 27, 1943, p. 8. "C'est le Boche! Mixing Beer, Burgundy!," *Training News*, Feb 19, 1944, p.7. "Unconquerables at Tuscaloosa," *Training News*, Jun 17, 1944, pp. 5, 11. "French Cadets Begin Training at Gunter Field," Montgomery Advertiser, Sep 3, 1943, p. 8.
136. AAF EFTC, 1 Jan 1943 – 31 Dec 1943, Vol. 2, pp. 1017-18, AFHRA Call # 222.01 1 Jan 1943-31 Dec 1943 v. 2, Doc # 00146142. "French Cadets At Tuscaloosa Find Life Here 'Le Paradis'," Training News, Aug 14, 1943, p. 7. Because of problems at the contract school at Tuscaloosa, the AAF decided to terminate primary training at this school at the end of July 1944 and transfer primary training of French cadets to the Hawthorne School of Aeronautics, Orangeburg, South Carolina, starting on August 7, 1944. AAF EFTC, 1 Jul 1944 – 31 Aug 1944, vol. 1, pp. 294-97, AFHRA Call # 220.01 AAF EFTC 1 Jul 1944-31 Aug 1944 v. 2, Doc # 00146152. "Unconquerables at Tuscaloosa," *Training News*, Jun 17, 1944, p. 11.
137. *Gunter Field History 1 Jan 1943 to 31 Jan 1944*, p. 119.
138. *Gunter Field History, 1 Jan 1943 to 31 Jan 1944*, pp. 68-70.
139. Letter, Col L. S. Moseley, Asst Chief of Staff A-3, AAF Flying Training Command, "French Speaking Flying Instructor," Nov 22, 1943, AFRHA Folder 353 French (1943), Call # 220.7272-2, Vol. 1 1943, Doc # 0014964.
140. "French Cadets Begin Training at Gunter Field," Montgomery *Advertiser*, Sep 3, 1943, pp. 1, 8.
141. "French Cadets Go To Gunter Field For Basic Training," *Training News*, Sep 4, 1943, pp. 1. 4.
142. "First French Cadets Leave Gunter; Set Fine Example For Other Classes," *Training News*, Nov 27, 1943, p. 9.
143. See Table 4.4 for FFAF statistics for the Gunter Basic School.
144. *AAF EFTC, 1 Jan 1943 – 31 Dec 1943*, Vol. 2, p. 102, AFHRA Call # 222.01 AAF

EFTC 1 Jan 1943-31 Dec 1943 v. 2, Doc # 00146142. *Gunter Field History*, 1 Jan 1943 to 31 Jan 1944, p. 121.

145. *AAF EFTC History, 1 Jan 1943 – 31 Dec 1943*, pp. 1026-29, AFHRA Call # 222.01 AAF EFTC 1 Jan 1943-31 Dec 1943 v. 2, Doc # 00146142. Interview by 1Lt Leland E. Hess, Historical Branch, Headquarters (HQ), AAF EFTC, with Captain Robert L. Lamaison, French Liaison Officer, HQ AAF EFTC, Apr 17, 1945, Tab 14, *AAF EFTC History, 1 Jan – 28 Feb 1945*, Vol. 2, AFHRA Call # 222.01 Jan-Feb 1945 v. 2, Doc # 00146161.

146. Letter, Col D. D. Fitzgerald, Chief of Staff to Gen Arnold, "French Training Conference, Jan 19, 1944, Chapter 27 Tab, *AAF EFTC History 1 Jan 1944 – 31 Jun 1944*, Vol. 4, AFHRA Call # 222.01 1 Jan 1944 – 31 Jun 1944 v.. 4, Doc # 00146148.

147. *AAF EFTC, 1 Jan 1944 – 31 Jun 1944*, Vol. 2, pp. 637–39, 644–47, AFHRA Call # 222.01 v. 2, Doc # 00146148. *Gunter Field History, 1 Feb 1944 to 30 Jun 1944*, pp. 48–51.

148. The Gunter basic school classes, Class 44-E, entry date of January 17, 1944, through Class 45-B, entry date of November 23, 1944, had a total of 3,247 students (2,513 Americans and 734 French trainees) (data taken from Table 9, Statistics for the Gunter Basic School). The Americans made up 77.4 percent of these classes, and the French made up 22.6 percent, approximately the same percentage as the American-to-French flight accidents during 1944. Data from Table 9. During the period Jun 1943 and Nov 1945, 20 French cadets died from training accidents and other causes at EFTC schools. The EFTC made arrangements to bury the deceased French cadets in Oakwood Cemetery next to the deceased RAF cadets. In the early 1980s, the French government through the French Consulate in New Orleans, LA, made arrangements to pay the Commonwealth (formerly Imperial) War Graves Commission (CWGC) for an extension to its maintenance contract with a private Alabama company for the RAF gravesites that would also provide for regular maintenance of the 20 French graves in Oakwood Cemetery. Training Command Memorandum 50–0-8, *Training General, French Training*, Apr 4, 1944, p. 2, AFHRA Folder Call #220.7272–1 1944, Doc # 00145963. Memo, T. F. Penfold, Chief Finance Officer, CWGC, "Montgomery (Oakwood) Cemetery Annexe, Alabama, USA Maintenance of French Graves, Nov 11, 1983; Folder "Montgomery Oakwood Cemetery," AU/HO Document Files, Maxwell AFB, AL.

149. *AAF EFTC History, 1 Jan 1944 – 31 Jun 1944*, Vol. 2, pp. 637–39, 648–49, AFHRA Call # 222.01 1 Jan 1944 – 31 Jun 1944 v. 2, Doc # 00146148.

150. *Ibid.*, pp. 644–47. Memorandum, Capt Robert L. Lamaison, French Liaison Officer, HQ AAFEFTC, to Col Paul E. Toda, [Dec 21, 1944], Chapter 16 Tab, *AAF EFTC History, 1 Nov 1944–31 Dec 1944*, AHRA Call # 222.01 Nov-Dec 1944, Doc # 00146159.

151. *AAF EFTC History, 1 Jul 1944 – 31 Aug 1944*, Vol. 1, pp. 292-93, AFHRA Call # 220.01 1 Jul 1944 – 31 Aug 1944, v. 1, Doc # 00146152.

152. *Maxwell Field History, 1 Nov 1944 - 31 Dec 1944*, p. 25, AFHRA Call # 286.26-5, Doc # 00175387.

153. *AAF EFTC, 1 Nov 1944 – 31 Dec 1944*, Vol. 2, pp. 417-28, AFHRA Call # 220.01 1 Nov 1944 – 31 Dec 1944, v. 2, Doc # 00146158.

154. *Maxwell Field History, 1 Jan 1945 to 28 Feb 1945*, p. 92, AFHRA Call # 286.26-6, Jan 1945 – Feb 45, Doc # 00175388.

155. *Maxwell Field History 1 Mar 1945 to 30 Apr 1945*, pp. 56-57, AFHRA Call # 286.26-7 Mar-Apr 45, Doc # 00175390.

156. *Maxwell Field History, 1 May 1945 to 30 Jun 1945*, p. 94, AFHRA Call # 286.26-8 May-Jun 45, Doc # 00175391.

157. *AAF EFTC History, 1 Jan 1945 – 28 Feb 1945*, Vol. 1, pp. 320-24, 417-28, AFHRA Call # 220.01 1 Jan 1945 – 28 Feb 1945 v. 1, Doc # 00146160.

158. *AAF EFTC History, 1 Sep 1944 – 30 Oct 1944*, Vol. 1, pp. 412–1, AFHRA Call # 220.01 1 Sep 1944 – 30 Oct 1944 v. 1, Doc # 00146154.

159. *AAF EFTC History, 1 Nov 1944 – 31 Dec 1944*, Vol. 1, pp. 407–09, AFHRA Call # 220.01 1 Nov 1944 – 31 Dec 1944 v. 1, Doc # 00146160. *Gunter Field History, 1 February 1944 to 30 June 1944*, p. 52.

160. *AAF EFTC History, 1 Mar 1945 – 30 Apr1945*, Vol. 1, pp. 340–43, AFHRA Call # 220.01 1 Mar 1945 – 30 Apr1945 v. 1, Doc # 00146162.

161. *Ibid.*, pp. 340–43. *Gunter Field History, 1 Mar 1945 to 30 Apr 1945*, pp. 18–21, AFHRA Call # 283.81–8 Mar-Apr 45, Doc #00172972.

162. *EFTC Year Book, A Summary Report of Training, Operations, and Activities AAF Eastern Flying Training Command for the Year 1944* (Maxwell Field, AL: Statistical Control Unit, HQ AAF EFTC, 1945), p. 7, in Exhibits, AFHRA Folder Call # 222.01 1 Nov – 31 Dec 1944 V. 3, Doc #00146159

163. *Gunter Field History, 1 Mar 1945 to 30 Apr 1945*, pp. 21–22, AFHRA Call # 283.81–8 Mar-Apr 45, Doc #00172972. "French Cadets Finish Training At Gunter Field," Montgomery *Advertiser*, Nov 24, 1943.

164. *Gunter Field History, 1 Jan 1945 to 28 Feb 1945*, pp. 12–13, 24, AFHRA Call # 283.81–7 Jan-Feb 45, Doc #00172971. *Gunter Field History, 1 Mar1945 - 30 Apr 1945*, p. 20, AFHRA Call # 283.81–8 Mar-Apr 45, Doc #00172972. "ATs Now In Use At Basic Schools," *Training News*, Jul 8, 1944, p. 1.

165. *Gunter Field History, 1 Jul 1945 – 31 Aug1945*, pp. 7, 9–11, AFHRA Call # 283.81–10 Jul-Aug 45, Doc #00172974.

166. Vigneras, p. 212. See p. 206 for Free French Air Force squadrons in December 1943.

167. White, pp. 38–39.

168. *Ibid.* In November 1943, the FFAF had a strength of 34,302 of which 1,778 were flying officers. Between June 1943 and January 1946, the AAF schools had trained 1,678 flying officers. The FFAF had lost 1,000 aircrew members from Jun 1940 to May 1945. From these numbers, the AAF-trained French flying officers made up about one-half of the FFAF flying officers in summer 1945. Christienne and Lissarague, pp. 398, 408

169. Vigneras, pp. 213. See Christienne and Lissarague, pp. 407–08, for wartime statistics of the FFAF. General Arnold was not so favorable toward the FFAF. He implied that the French were "playing politics" in submitting a specific request for aircraft and that the Free French military forces had contributed little to the war effort, despite the "vast quantities of equipment and supplies they received." Arnold, pp. 541, 543.

170. "Unconquerables at Tuscaloosa," *Training News*, Jun 17, 1944, p. 11.

171. Hines, p. 37.

172. See Hines, Chapters IV-VI, pp. 45–98, for narrative about US Air Force flight training programs from 1950 to 1976.

173. AETC/HO, "Major Changes in Undergraduate Pilot Training 1939–1998," at http://www.aetc.af.mil/ Shared/media/document/AFD-07130–081.pdf, accessed May 22, 2013. AETC/HO, "Specialized Undergraduate Pilot Training Pilot Production FY92-F"Y02, at http://www.aetc.af.mil/pilot/productprodction/Chart.htm,accessed

May 22, 2013.

174. 49th Fighter Wing Public Affairs Office, "German Air Force," Jul 14, 2008, at http://www.holloman.af. mil/library/factsheets/factsheet.asp?id=4364.

175. 82nd Flying Training Wing (FTW) Public Affairs Office, "Euro-NATO Joint Jet Pilot Training Program (ENJJPT)," Oct 17, 2012, at http://www.sheppard.af.mil/library/factsheetspage/factsheet.asp?fsID=5168. William M. Clifton, 80 FTW Historian, "Background Paper on International Undergraduate Pilot Training at Sheppard Air Force Base, 1966–2013," Jun 11, 2013; provided to author by Mr. Clifton. E-mail, Lt Col Marc J. Himelhoch, US Senior National Representative, 80 FTW, Sheppard AFB, TX, to Robert B. Kane, "Euro-NATO Joint Jet Pilot Training Program," Aug 6, 2013.

176. James Lawton Collins, Brig Gen, US Army, *Vietnam Studies: The Development and Training of The South Vietnamese Army, 1950–1972* (Washington, DC: US Army, 1991), pp. 112–14. "Laos – A Secret Pilot War," Mar 12, 2011, at http://heritage-flightgeardisplays.wordpress.com/2011/03/12/laos-a-secret-pilot-war/, accessed Aug 8, 2013.

177. Interview, author with Mr. Gregory L. Baker, Foreign Disclosure Officer, Air University, Maxwell AFB, AL, Aug 14, 2013. "Defense Language Institute English Language Center," n. d., at http://www.dlielc.edu/, accessed Aug 15, 2013.

178. See Hines, pp. 16–17, 23, 58–65, 85–89.

179. *Ibid.* International Officer School History, n.d., provided by Mr. Gregory Baker, IOS (cited hereafter as IOS history). Author's personal knowledge. I served as the Chief of Instruction for the International Officer School (IOS) from July 2001 to July 2003 and managed the three preparatory courses for the international students who attended the Air University PME schools at Maxwell AFB. In addition to conducting the preparatory courses and the sponsor program, the IOS also provides the spouses and older children of the international officers with six weeks of conversational English instruction and provides other administrative services for the officers and families during their time at Maxwell AFB.

180. *Ibid.* IOS history. Author's personal knowledge as my family also serves as sponsors for the Indonesian students and families who have attended Air University schools since July 2003.

181. Jean Martin, "The Golden Years: John Clark," Selma *Times-Journal*, Oct 16, 2010, at http://www.selmatimesjournal.com.2010/10/16/the-golden-years-john-clark/, accessed Jun 6, 2014. "Obituary John Clark," http://obits.al.com/obituaries/birmingham/ obituary.aspx?pid=167605659, accessed Jun 6, 2014. Quinn, p. 320. Quinn also provides a short narrative (pp. 550–59) of Miss Harmon's story of her encounter and marriage to John Clark and her time in Britain during the war.

Index